Using the Maple V Flight Manual with Release 2 or Release 3 of Maple.

To use the Maple V Flight Manual with **Release 2** *or* **Release 3**, make these changes:

Page 48

The plot( ) command at the bottom of the page will generate an "empty plot" warning from Iris. To get the correct plot, change it

from: `plot(f, -5..5);`   to: `plot(f, x = -5..5);`

Page 94

The plot command as the top of the page will not generate an Iris error message, but it will not plot as intended. To get the correct plot, change it

from: `plot(p, x=0..6);`   to: `plot(p(x), x=0..6);`

To use the Maple V Flight Manual with **Release 3 only**, make these changes:

Page 16

In between the two lines that start with the command isolate(", ...); issue this command:

`convert(", radical);`

Using Release 3 here, the " should refer to the output `x - 5/6 = RootOf(36*_Z^2 - 13)` and this convert command should yield `x - 5/6 = 1/36 * 13^(1/2) * 36^(1/2)`.

Page 53

Replace the `powsubs( )` command at the top of page 53 with this command:

`simplify(", symbolic);`

Notice that, in the sequence of operations in this section, you are replacing the second of two uses of the powsubs( ) command. Leave the first one as given, and just replace the second one.

RootOf (The "generic" explanation for the previous two recommendations.)

Release 3 is somewhat more "conservative" than earlier releases about explicitly calculating roots. Thus, R3 will yield answers of the form RootOf(expr) more often than did earlier versions. You may well see RootOf results in other places when you try to perform similar operations on your own. Another tool for working with RootOf results is:

`s := allvalues(..., d);`

where the use of the d-option helps to simplify Maple's task, and using a name such as s means you can then work with distinct solutions, s[1], s[2], ...

Page 60

To modify 2D or 3D Plot Styles in Release 3 for the Macintosh, click on the **Options** button in the Plot Window. When you are done making your choices, click on **Apply** in the Plot Options window and the plot will be redrawn. In Release 3 for DOS/Windows, options may be modified via menus in the Plot Window itself. When done, in Windows, click on the right mouse button to redraw the plot. In DOS, **Escape** to the main plot screen and press **Enter** to redraw the plot.

The 2D and 3D Plot Options are further described in the *Getting Started* book for your platform, and the Maple Help files for plot[options] and plot3d[options].

Chapter 4, Linear Algebra

In the linear algebra section, matrices are typically given names consisting of a single upper-case letter. While that is consistent with common practice in linear algebra, it is inconsistent with Maple's use of D for the differential operator, E for the base of the natural logarithms, and I for the complex number representing the square root of negative 1.

Prior to Release 3, Maple would let you use such names for matrices. In doing so, you would just overwrite Maple's meaning for them. Starting with Release 3, however, Maple protects the names it uses. So, for example, the use of D := {seq(...)}; on page 97 will generate an error message in Release 3. There are two ways around this. You must use one of these in Release 3. You may use either one in earlier releases if you want, it is just not required with them.

The best way is simply to substitute a letter other than D (but not E or I) every time D is used to refer to a matrix. This will leave the differential operator "D" in place.

Since that can be rather tedious to remember, if you will not need the differential operator "D" during your Maple session, then you can alias it to something else, such as DD. This will unprotect it, and Maple will then let you use D (or DD) as the name of a matrix. You can implement this second strategy by extending the recommended alias command (introduced on page 81 and repeated on pages 87, 96, 104, and 110) to:

```
alias(I=&*(), D=DD);
```

Chapter 5, Differential Equations

Many of the activities in this chapter require routines that are not built in to the basic Maple distribution. Prior to Release 2, these routines were distributed in a file that, from various sources, was called ODE or ODE.m. Throughout this chapter (e.g., pages 118, 127, 137), you are instructed that you must have read the ODE (or ODE.m) file at some point in your Maple session before you can continue with the activities.

Beginning with Release 2, the routines contained in the ODE file have been distributed with the Maple "share" library and, on page 117, you are given instructions on one way to create your own ODE.m file. The method for accessing the share library has changed with Release 3 so, if you still want to use that method, you must replace the three input commands on page 117 with:

```
with(share);
readshare(ODE,plots);
save `ODE.m`;
```

You need to do this only once, and you will have created your own copy of the ODE.m file. After that, you will simply need to issue the command read `ODE.m`; at the beginning of each Maple V session (or in your MapleInit (Mac) or maple.ini (DOS/Windows) file).

The above technique, however, creates a larger-than-necessary ODE file. An alternative approach is to not bother creating a separate ODE file to read, but to go directly to the share library in each session where you need these routines. In that case, every time you are instructed to read the ODE file, if you have not already done so in your current Maple session, you can issue the following two commands instead:

```
with(share);
readshare(ODE,plots);
```

As with the first strategy outlined above, you can issue these commands automatically via an initialization file, or manually within your Worksheet.

Maple® V Flight Manual

Tutorials for Calculus,
Linear Algebra, and
Differential Equations

Maple® V Flight Manual

Tutorials for Calculus, Linear Algebra, and Differential Equations

Wade Ellis, Jr.
West Valley College

Eugene W. Johnson
University of Iowa

Ed Lodi
West Valley College

Daniel Schwalbe
St. Cloud State University

Brooks/Cole Publishing Company
Pacific Grove, California

Brooks/Cole Publishing Company
A Division of International Thomson Publishing Inc.

Printed in the United States of America

10   9   8

Library of Congress Cataloging-in-Publication Data

Maple V flight manual : tutorials for calculus, linear algebra, and
    differential equations / Wade Ellis, Jr. . . . [et al.].
        p.   cm.
    Includes index.
    ISBN 0-534-21235-2
    1. Maple (Computer program)   I. Ellis, Wade.
QA155.7.E4M35   1991   91-36001
510'.285'5369—dc20   CIP

Customer support is available to registered users of Maple between 9 a.m. and 4 p.m. Pacific Time at (408) 373-0728, or E-mail via Support@BrooksCole.Com

Sponsoring Editor: *Jeremy Hayhurst*
Editoral Associate: *Nancy Champlin*
Production Coordinator: *Marlene Thom*
Manuscript Editor: *Timothy Phillips*
Interior Design: *Vernon Boes*
Cover Design: *Lisa Berman*
Art Coordinator: *Cloyce J. Wall*
Compositor: *the Computer Tutors of San Jose*
Printing and Binding: *Malloy Lithographing, Inc.*

Preface

This book updates and expands an earlier work entitled *Maple® for the Calculus Student: A Tutorial*. The new version of Maple®, called Maple V, is a major software revision that includes both enhancements to the user environment and extensions to the set of Maple commands. The short linear algebra and differential equations chapters of the previous work have been expanded into major chapters in this book and incorporate the many extensions of Maple in these areas. The increased speed of two-dimensional graphing and the addition of three-dimensional graphing markedly improves the visualization possibilities of Maple. These new capabilities are included in this book.

The first three chapters are primarily the work of Ed Lodi and Wade Ellis, Jr., although the two other authors made many contributions to these chapters as well. The fourth chapter, on linear algebra, is the work of Eugene Johnson. His efforts also include extensive consultation with the Waterloo Maple Software Group (WMSG) in extending the Maple `linalg` package. Dan Schwalbe created the fifth chapter on differential equations with a great deal of editing help from his wife Kathryn Schwalbe. He also wrote the set of differential equations commands contained in the **ODE** file.

The authors of this book have greatly enjoyed working together on this project for more than a year. The sharing of ideas about computer algebra systems and how to use them in doing, learning, and teaching mathematics has been immensely rewarding. We only hope that the reader will enjoy this work as much as we, the authors, enjoyed the labor of creating it.

Acknowledgments

We would like to express our thanks to the following people from Brooks/Cole Publishing Company: Jeremy Hayhurst and Robert Evans for their ongoing efforts toward making Maple accessible to students in mathematics courses and for bringing the authors together on this project; Marlene Thom and Nancy Champlin for their helpfulness and good cheer throughout the production of this book; Timothy Phillips for copyediting the manuscript; and the design and production staffs, especially Vernon Boes, Lisa Berman, and Cloyce J. Wall, for their many helpful suggestions. We are pleased to have worked with the outstanding staff at Brooks/Cole Publishing Company.

We would like to thank the Waterloo Maple Software Group for expanding and refining the Maple V version of the **linalg** package so that it would be more powerful and easier to use. We especially thank WMSG's Mike Monagan, who was the principal programmer for the **linalg** package. The many exchanges of ideas and code with Mike enhanced Maple's **linalg** package, Dan's set of **ODE** commands, and the chapters on linear algebra and differential equations. Thanks go to Keith Geddes for initiating and supporting the decision to modify Maple V so that it would be more useful in an instructional environment. We thank George Labahn for his enthusiasm, ideas, and sharing his code, parts of which helped speed the development of the commands in the **ODE** file. Also, the technique of putting arrows on vectors is his. We thank Greg Fee for implementing many of the improvements in the **linalg** package. We thank Benton Leong for facilitating the timely approval and implementation of software changes and Janet Cater for her help in working with Maple V on the Sun computer. Others in the Waterloo group have also been helpful in testing and making suggestions.

We also thank students in the University of Waterloo Symbolic Computation Group, especially Blair Madore, Lee Qiao, and Katy Simonsen.

Our thanks go to Sun Microsystems, Inc., for the loan of a computer to run Maple V in the Sun environment and especially Shernaz Daver and Eva Markett for their help in using the Sun Workstation.

Many thanks also go to William C. Bauldry of Appalachian State University, Maurino P. Bautista of Rochester Institute of Technology, Sharad Keny of Whittier College, Glenn Sowell of the University of Nebraska at Omaha, and Jeanette R. Palmiter of Portland State University for reviewing the manuscript and for their many suggestions for improving the book. Of course, any errors that remain are our own.

Finally, we would like to thank our wives—Jane, Sandy, Rose, and Kathryn—for their patience, support, and encouragement during the writing and production of this book.

Wade Ellis, Jr.
Eugene W. Johnson
Ed Lodi
Daniel Schwalbe

How to Get the ODE File for Non-Macintosh Computers

The routines used in chapter 5 of the *Maple V Flight Manual* are contained in a file called **ODE**. This file can be obtained from the Maple Share Library or by writing to Brooks/Cole Publishing Co., 511 Forest Lodge Road, Pacific Grove, CA 93950-5098, Attn: Heidi Wieland.

There are two ways to retrieve the file from the Maple Share Library—through Internet's anonymous file transfer protocol or through Bitnet's e-mail.

Anonymous FTP

If you have access to ftp distribution, you can log into the Maple server yourself. After logging into your ftp-capable system, type the following:

ftp daisy.waterloo.edu

If you get a login prompt, type in the userid

anonymous

and the empty (null) password.

Your screen should then look something like the following:

```
%ftp daisy.waterloo.edu
Connected to daisy.waterloo.edu.
220 daisy FTP server
(Version 5.103 Sat Nov 17 21:57:34 EST 1990) ready.
Name (daisy.waterloo.edu:schwalbe): anonymous
331 Guest login ok, send ident as password.
Password:
230 Guest login ok, access restrictions apply.
```

You are now logged into the server. You may use the commands **ls** to list a directory's contents and **cd** to change directories. Note that **cd /** will return you to the root directory. The command **get** is used to retrieve a file to your system. To retrieve the **ODE** file, issue the following commands:

cd /maple/5.0/share/plots
get ODE

To stop the ftp session, type **quit** at the prompt. There are also several informational files in **/maple** that you may want to retrieve if you have further interest in Maple Share Library. Type **/maple** to list these files. Specific files of interest include **guide**, **help**, **README**, and **address**.

E-Mail

If you do not have access to anonymous ftp, you can use e-mail to retrieve the file. The e-mail address of the Maple Share Library is

maple-netlib@can.nl

Send an e-mail message to this address with the following message:

send plots.ODE from share5.0

The file is currently stored as **plots.fieldplot** until the next update to the library.

To learn more about the Maple Share Library, send the following command to the above address.

send index

You may have to consult an expert to transfer the file from your e-mail system to the machine on which you are using Maple.

Contents

Introduction   xv

What are Computer Algebra Systems?   xv

The Origins of Computer Algebra Systems   xv

The Origins of Maple   xvi

How to Use This Book   xvi

A Note to Users of Mathematics   xvii

Introduction for Instructors   xvii

Computer Algebra and the Mathematics Curriculum   xvii

The Purpose of the Book   xvii

The Structure of the Book   xviii

Ways This Book Can be Used   xx

1 Getting Started   1

Maple and Mathematics   3

Using the Help Feature   3

Learning to Use Maple   4

Numbers: Integers, Rationals, Decimals   5

Variables   7

Editing on the Macintosh   10

Editing on the IBM   11

Extensions   11

Help Facility   11

Macintosh Environment   12

IBM Environment Under Windows   12

IBM Environment Under DOS   13

Additional Activities   13

Entering Expressions   13

Exploring Expressions   14

2

Precalculus   15

2.1   **Solving Equations**   **15**
Polynomial Equations   15
Other Types of Equations   18
Inequalities   19
Additional Activities   21

2.2   **Rational Expressions**   **21**
Additional Activities   23

2.3   **Defining Functions and Procedures**   **24**
Functions of One-Variable   24
Functions of Several Variables   26
Automating Commands in Maple   26
An Automating Procedure   27
Automated Plotting   28
Extensions   29
   Defining Functions   29
   Automating Commands   29
Additional Activities   31

2.4   **Additional Features of the `plot` Function**   **31**
Parametric Form   32
Polar Form   33

2.5   **Putting it All Together**   **34**
A Rational Function with Asymptotes   34
Finding the Roots of Polynomials   36
Additional Activities   39

2.6   **Matrices and Matrix Commands**   **40**
Additional Activities   42

3

Calculus   45

3.1   **Calculus I**   **45**
Limits of Functions   45
Derivatives of Functions   48
Investigating a Function Using `plot`
     and `diff`   48
Integrals of Functions   50

Definite Integrals of Functions of
One-Variable   51
Investigating a Practical Problem   53

3.2   **Calculus II   56**
Series   56
Taylor Series   56
Functions of Several Variables   57
Derivatives of Functions of Several
Variables   57
Applications of Partial Derivatives   58

3.3   **3D Plotting and Functions of Several Variables   59**
Graphing Functions of Two Variables   59
Multiple Integrals of Functions of Several
Variables   61
An Application Using Multiple Integrals   61
Limits of a Function of Several Variables   62
Extensions   62
Maple Calculus Commands   62
The Student Package   63
Additional Activities   63

4

Linear Algebra   67
The linalg Package   67

4.1   **Matrices and Vectors   67**
Solving Systems of Linear Equations   69
Further Editing Commands   74
An Important Point on Names   75
Summary and Extensions   75
Additional Activities   76

4.2   **More on Matrices and Vectors   77**
Matrix and Vector Arithmetic   77
Other Forms of the Matrix and Vector Commands   81
A Note on Arrays, Loops and Sequences   83
Summary   85
Additional Activities   85

4.3   **Basic Matrix and Vector Functions   87**
More on Subscripts and Sequences   90
Polynomial & Matrices: The Cayley-Hamilton Theorem   91

Polynomial & Matrices: Curve Fitting   92
Summary and Extensions   94
Additional Activities   94

4.4   **Maple's Basis and Dimension Commands**   **96**
Sums and Intersections of Subspaces   97
Summary and Extensions   98
Additional Activities   98

4.5   **Linear Transformations**   **99**
Orthogonal Projection   101
Summary   102
Additional Activities   103

4.6   **Eigenvalues and Eigenvectors**   **104**
Summary   108
Additional Activities   109

4.7   **Diagonalization and Similarity**   **109**
Diagonization   110
Similarity and Smith Form   113
Similarity and Frobenius Form   114
Summary   115
Additional Activities   116

5

Differential Equations   117

5.1   Introduction   117
First Order Differential Equations   118
Using fsolve in an Application   121
Verifying Solutions   124
Additional Activities   125

5.2   Graphical Procedures   126
Direction Fields   126
Additional Activities   131

5.3   Numerical Procedures   132
Euler's Method   132
Additional Activities   136
Improved Euler   136

Runge/Kutta 137
Additional Activities 141

5.4 Picard Iterates 141
Additional Activities 143

5.5 Higher Order Equations and Systems 144
Homogeneous Linear Systems with Constant Coefficients 146
Additional Activities 153

5.6 Phase Space 154
Additional Activities 159

5.7 Numerical Methods for Systems 159
Runge/Kutta 159
Additional Activities 162

5.8 dsolve Options 164
LaPlace Transforms 164
Heaviside Function 166
Additional Activities 168
Series Solutions 169
Recurrence Relations 171
Bessel Equations 174
Additional Activities 175

Appendix A 177

Index 179

Introduction

What are Computer Algebra Systems?

You have learned to solve equations and explore functions in your previous mathematics courses. The sophisticated manipulation of symbols and expressions that you use to solve equations and investigate functions can also be performed by software packages called computer algebra systems. You can use computer algebra systems to generate the exact symbolic solutions you determined by hand, the numerical approximations you found with a calculator, and the graphs you have drawn. As with paper-and-pencil methods of solution and finding graphs, however, the success of computer methods are limited by available time, space, and ability.

The Origins of Computer Algebra Systems

In 1959 at the Massachusetts Institute of Technology, a group of researchers began the development of MACSYMA, a computer algebra system. This first system was the outgrowth of an attempt to convince the science community that computers could perform significant intellectual tasks. Mathematics was chosen as the vehicle for demonstrating the intellectual possibilities of machines since some important and difficult mathematical processes are rule-based, highly structured, and, thus, possibly programmable on a computer. MACSYMA's early successes proved that the computer programming of manipulative mathematics (like the solving of many elementary differential equations) was not nearly as difficult as one might assume given how hard it seems to be to teach these procedures to people.

The Origins of Maple

In 1980 a group of professors and researchers at the University of Waterloo in Canada began discussions of computer algebra systems and their use in engineering and various fields of mathematics. They decided that the currently available systems were not appropriate tools for the needs of their university with its strong engineering and mathematics departments where both research and teaching are important. The Symbolic Computation Group was formed and decided to develop a new computer algebra system that would satisfy several criteria: (1) It should allow many users (including students) to access the system at the same time, requiring minimal computer processing power or memory; (2) it should have a clearly logical and understandable syntax; and (3) it should allow for additional features to be added easily. The outcome of this effort was called Maple.

Maple has been extended and enhanced many times in the last decade. Maple V, the current release for which this book is written, embodies changes to the graphing capabilities of Maple to include three-dimensional graphing of real-valued functions of two variables. Many new commands and functions, especially in linear algebra and differential equations, have been added as well. Maple V also allows for easier function definition and performs many computations faster than previous releases. This new implementation includes improved speed in two-dimensional graphing, and (on appropriate hardware) color graphics capabilities and more flexible text formatting.

How to Use This Book

You can obtain the greatest benefit from this book by thoughtfully working through each item. You should carefully consider the commands you are asked to enter and think about the expected result before you enter the command. Occasionally, in the early chapters of the book, you will be asked to enter commands that will give rise to error messages. This is to provide experience with such errors and to encourage you to experiment fearlessly with Maple commands.

A Note to Users of Mathematics

Maple V is a powerful mathematical tool that has already found significant uses in investigating and analyzing problems in the natural and managerial sciences, and in engineering. The symbolic, numeric, and graphical power that Maple V provides can greatly assist you in the mathematical aspects of your work. This book quickly familiarizes you with the Maple V environment and some of the many ways of using Maple in solving problems involving calculus, linear algebra, and differential equations.

Introduction for Instructors

Computer Algebra and the Mathematics Curriculum

The mathematics curriculum can be enriched through the intelligent use of computer algebra systems, but this enrichment is not without cost. There will be some trade-offs. The courses will change to reflect the new technology which will result in a new arrangement of course priorities. Instructors will need to spend more time in teaching these enriched and modified courses. Such new courses should be more interesting, exciting, and challenging for both students and faculty. Students with access to a computer algebra system will be able to devote more time to concepts and applications. They can actually experiment with mathematics in a way that was highly impractical before. Thus, computer algebra allows students to sharpen high-level problem-solving skills while relying on the computer to handle the more mundane details.

The Purpose of the Book

The first three chapters of this book are based on material from precalculus and calculus and are intended as an introduction to the use of computer algebra systems in mathematics and mathematics courses. Although the first three chapters can be used as a supplement to standard calculus courses, a more thorough integration of computer algebra into such courses can best be achieved with materials from the books like *Calculus Laboratories for Brooks/Cole Software Tools; A Tool, Not an Oracle: Maple Laboratories for the Calculus Student*; or *Maple V Calculus Labs*.

Chapter 4 provides an outline for a theoretical first course in linear algebra that assumes that any extended computation will be performed using a computer algebra system. Chapter 5 provides an outline for a differential equations course that hinges on computer-based graphing, symbol manipulation, and graphics. Most differential equations are investigated from a graphical and numerical point of view since such equations most often do not possess closed-form solutions. The first three chapters, including a brief introduction to Maple V programming, provide the necessary foundation for the mathematical work in the last two chapters.

The Structure of the Book

This book, as indicated, is constructed around the standard course content of the first two-years of college mathematics including the calculus of one- and two-variables, linear algebra, and differential equations. The first two chapters introduce well-understood precalculus topics as "test" data for mathematical experiments and investigations with the various Maple commands or operators. Once the reader is familiar with Maple's command structure, help facilities, and error recovery methods, the Maple calculus functions are introduced and techniques for using sets of Maple functions to attack mathematical problems are presented in Chapter 3. The presentation of two-variable calculus topics in this book assumes greater sophistication of the reader; and, therefore, the tutorial features of the book occur less frequently. There is a brief introduction to the automation of sets of commands (also known as programs) at the end of this chapter.

The first three chapters are an introduction to the use of Maple V as a supplement to the standard techniques of paper-and-pencil computation learned in calculus. Though other approaches to this material are possible in light of computer algebra systems, we have tried a minimalist approach in these introductory chapters with only a brief discussion of programming and no required programming.

In the linear algebra chapter, Maple is used from the outset to restructure the standard topics from a conceptual point of view rather than a computational one. This chapter is not a textbook for a linear algebra course, but it does provide the necessary computational techniques and rationale for the use of computer algebra systems as an integral part of an introductory linear algebra course. Involved or lengthy computations are done from the beginning with Maple rather than with paper-and-pencil methods allowing greater time to develop and illuminate the conceptual framework behind the computations. Readers who have previously taken courses that use the first part of this book, will feel comfortable in working through the linear algebra chapter even though there are fewer tutorial aspects of the presentation. Once again, an ever-increasing sophistication in mathematical thinking and computer use is assumed and expected of the reader. The reader is often asked to enter several Maple commands at once and to reflect more deeply on the results computed by Maple.

In Chapter 5, a new tone and direction for the study of elementary differential equations is developed based on the complete array of tools found in Maple. The graphical, numerical, and symbolic capabilities of Maple are used to lay the groundwork for a course in which modeling becomes the central theme of the course. The power of Maple allows one to experiment with and become familiar with more interesting examples which motivate the study of the underlying mathematics. Getting an "answer" is no longer sufficient, understanding the model is the goal.

Readers will gain a more accurate view of the usefulness and beauty of differential equations by understanding more about the nature of solutions rather than concentrating on paper-and-pencil computational skills in solving specific types of elementary differential equations. Though this chapter is not a textbook on differential equations, it points the way toward a new, more relevant and interesting course. This new course is more demanding, but more stimulating and rewarding.

Ways This Book Can be Used

This book can be used as a computer supplement to any of the courses in the first two years of college. The introductory tutorial material on precalculus mathematics develops the necessary understanding of the use of Maple for the further use of Maple in calculus, linear algebra, or differential equations. Individuals who are using technology for the first time in teaching these courses may wish to begin with small technological enhancements. This book will assist such teachers in making incremental changes to courses while preparing themselves to make more substantial changes based on the instructional and intellectual mastery of a computer algebra system.

Because of its structure and varied emphasis, entire mathematics departments can use this book to gradually, but effectively, introduce computer algebra systems into their course offerings. The tutorial nature of the first chapters allows students and professors alike to become conversant with a computer algebra system without greatly overhauling the traditional courses. Once students and professors are aware of what computer algebra systems can and cannot do, then using them to perform computations in a linear algebra course becomes a simple extension of well-developed ideas about computer computations. Finally, with a sophisticated concept of how and when to use computer algebra systems most fruitfully, a completely revamped differential equations course becomes a natural outgrowth of the use of a computer algebra system. With an altered vision of the true nature of the differential equations course, professors will be more able to restructure the elementary calculus courses to better develop mathematical maturity in their students through the appropriate use of computer algebra systems.

Courses intended for the mathematical preparation of teachers can use this book as a supplement on computer usage for schools. The wide variety of mathematical content that can be addressed from the Maple package will give the prospective teacher a firm grounding in the possibilities of computer usage in teaching mathematics and the need for their students to become familiar with the uses and possible abuses of technology.

This book can also be used in computer courses intended for engineers and scientists with titles like Introduction to Computing for Engineers and Scientists. Though these courses are now often taught using the numerical capabilities of Fortran or Pascal, the future engineering and scientific use of computers will surely involve computer algebra systems. The book can be used either as a supplement or as the textbook for such a course.

1 Getting Started

Both the Macintosh environment and the IBM DOS and Windows environment will be introduced in this section. Most of the environments in which Maple can be used will be similar to one of these.

If you have a Macintosh computer, turn it on now. Use the mouse to open the Maple folder (if necessary) and the Maple application. Your screen should look something like this:

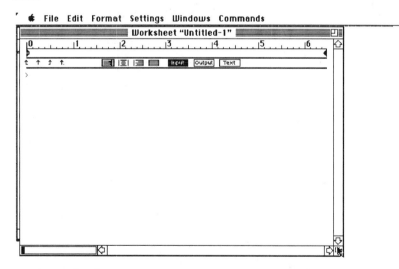

The **>** is the normal Macintosh prompt inviting you to enter a Maple command.

If you are using an IBM computer or compatible under Windows, start the the Maple application by first typing

win

and press Enter. Then double click on the Maple V icon to start the application. Your screen should look something like this.

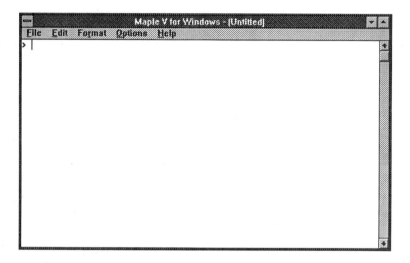

The arrowhead is the standard prompt for the IBM and compatible computers. Notice the five menu items at the top of the screen. These menus give you access to numerous activities and tasks you can execute when needed. For instance, you can use any one of three sizes for the mathematics displayed by selecting Math Style under the Format menu. Selection can be made by using either the mouse or pressing the key that is underlined.

If you are using an IBM computer or compatible under DOS, start the the Maple application by typing

maple

and pressing Enter. Most of the screen will be blank except for the menu items at the bottom of the screen. Your screen should look something like this:

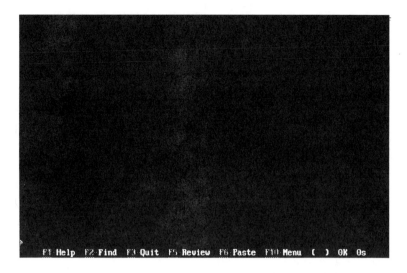

F1 Help   F2 Find   F3 Quit   F5 Review   F6 Paste   F10 Menu   ( )   OK   0s

Maple and Mathematics

You can easily learn to use the powerful capabilities of Maple if you thoughtfully work through the following tutorial. It is important, therefore, for you to enter each of the Maple statements on the right that appear in color. The slanted sentences on the left give you a description of the commands on the right. Additional explanation follows the Maple commands that appear in color.

You can maximize your understanding if you think about each of the Maple commands and the explanations that follow them rather than entering them in a mechanical fashion. The intent of this presentation is to give you a feel for the interaction that occurs between you, your computer, and Maple. After you are comfortable working with your computer environment and Maple, you will work through examples that will help you learn how to investigate mathematical problems and concepts using Maple.

Using the Help Feature

Use ? to get help in any environment.

?

After you type the ? symbol, press Enter on the Macintosh and IBM or press Return on the Sun to view the Help screen. This book will use "press Enter" to mean "press Enter or Return" as determined by your environment. This is a general help screen that describes how to get help on various topics. The description below the Synopsis provides additional information and explanation. For example, Note 3 indicates that commands must end with a semicolon (;).

You can use the Help facility to find and learn how to use Maple commands. You are encouraged to use it often.

To return to the Maple environment, click on the Go Away or Close box in the upper left corner (use the left mouse button on the IBM and Sun). Select Close under IBM Windows. Under DOS, you need to press the Esc key.

You can get help on a specific topic.

```
?intro
```
Remember to press Enter. This gives an introduction to Maple. You may wish to read this carefully or return to it later.

Learning to Use Maple

You can add two numbers.

```
2 + 3;
```
Notice the semicolon (;) at the end of the line. It is used as a terminator for each statement in Maple. After you have typed the semicolon, you should press Enter. (Remember to press Return on the Sun workstation.) The answer 5 appears in the middle of the screen followed by the appropriate prompt for your computer and the Maple V version you are using.

You can also add fractions.

```
2/3 + 1/7;
```
The fractions are typed with a division sign (/). After you type the semicolon (;), you should press Enter. Notice that the answer, $\frac{17}{21}$, is displayed in the center of the screen (with a fraction bar and the 17 directly above the 21).

A number can be raised to a power.

```
2^5;
```
The caret (^) indicates exponentiation. You have asked Maple to find 2^5. As you know, this is 32. Maple also

accepts the double asterisk operator (**) to indicate exponentiation.

What happens when you forget to type the semicolon (;) before you press Enter?

```
2 + 3
```

To simulate this error, be sure **not** to type the semicolon (;) before you press Enter. The cursor moves down one line. You may wait for some time before you discover that no answer has appeared. A semicolon (;) is missing at the end of the statement. Thus, Maple is waiting for you to end the statement.

You can correct this problem.

```
;
```

Type a semicolon and press Enter. The answer 5 appears in the center of the screen when you do this. The Maple prompt appears, inviting you to enter another statement. On the Macintosh, you can also use the Format menu's Go to Prompt item to obtain the > prompt should you lose it.

If your keyboard has only a Return key or only an Enter key, go on to the next section.

The following problem occurs frequently for beginning and experienced Macintosh users: pressing the Return key (a wide key) when you should press the Enter key (or vice versa on the Sun).

```
2 + 3;
```

To see this error, be sure to press the Return key instead of the Enter key (or vice versa). The cursor moves down to the next line. Nothing else happens on the screen. Pressing the correct key is one way to recover from this problem. You should do that now. This is not a problem on the IBM computer.

Numbers: Integers, Rationals, Decimals

You can also raise 2 to large powers.

```
2^32;
```

Notice that the answer is displayed exactly. This result is different (and correctly so) from what you would obtain on a calculator.

You can raise rational numbers to large powers as well.

```
(2/5)^32;
```

The parentheses are required. The fraction displayed is once again the exact answer.

A different result occurs if you use another representation of two-fifths.

`0.4^32;`

Notice that the displayed answer differs from the previous answer. How many digits are displayed? This answer is similar to what you would expect from a calculator and is the best 10-digit approximation to the exact answer.

You can force the exact rational answer to be approximated by a decimal using the Maple `evalf` *command.*

`evalf((2/5)^32);`

Take careful note of the parentheses. The outermost parentheses enclose the argument of the Maple command **evalf**. This result matches the result for the previous statement. It also has 10 digits.

Mistakes often occur when you are entering expressions that have many parentheses in them.

`evalf(2/5)^32);`

Be sure to enter the parentheses exactly as shown. The message displayed is a standard message that occurs when Maple does not understand what you entered. Here, there are more right parentheses than left parentheses. You should check for an incorrectly formed statement or a misspelled word whenever a syntax error message occurs. Notice that a ^ appears where Maple believes the error to be. You can either retype the statement or use the editing features of your computer to execute the desired statement.

You can specify the number of digits that will be displayed in decimal representations of numbers.

`Digits := 20;`

You must type a capital D in Digits since Maple differentiates uppercase and lowercase letters in names. The := symbol (with no space between the : and =) instructs Maple to assign the value on the right to the variable on the left. This increases the number of digits displayed from the standard 10 to 20. You see that the variable *Digits* is now 20 by the displayed result:

`Digits := 20.`

Let's look at the decimal representation of 2^{100}.

`evalf(2^100);`

Here the result is displayed in 20-digit decimal format. This is also called floating-point format.

You can vary the number of digits in a particular floating-point evaluation.

`evalf(2^100, 30);`

Notice that the answer now has thirty digits even though

Digits is set to 20. Reset *Digits* back to 10 with the command **Digits := 10;**.

Variables

You can assign a value to a variable.

`z := 5;`

This statement assigns the value 5 to the variable z. In previous interactions with Maple, you saw an answer. Here, Maple echoes the statement you entered (without the ending semicolon). Maple echoes the statement you enter when no operation is performed that results in an answer.

Maple can evaluate expressions.

`z^2;`

The variable z has the value 5. Maple assigns z this value and then evaluates the expression.

Multiplication must be explicitly indicated in the expression to be evaluated.

`2*z;`

Multiplication is indicated by an asterisk (*). The value of the expression when z is 5 is clearly 10.

What happens if you forget an asterisk (*)?

`2z;`

The syntax error message indicates that there is a problem with the statement as entered. You have to determine what correction(s) must be made. Notice the ^ under the **z**. You would need to reenter the statement (or correct it) with the missing multiplication symbol correctly placed.

Reestablish z as a variable with no value assigned to it.

`z := 'z';`

The ' symbol is on the same key with the quote (") symbol to the left of the Return key.

Check that z has no assigned value.

`z^2 + 4*z;`

The expression you entered is displayed in a different way. Notice that z squared is displayed in standard mathematical notation. Also, there is no asterisk in the displayed expression, but there is a space between 4 and z. While you *must* enter expressions with the exponentiation and multiplication symbols, Maple displays expressions without them.

The " symbol is assigned the value of the last expression evaluated.

```
";
```
The expression $z^2 + 4z$ is displayed in Maple format.

You can use the " in expressions.

```
" - 2*z;
```
Maple combines like terms when possible, as you can see.

You can change the expression with other algebraic operations.

```
" - 3;
```
Here, the -3 is simply appended to the expression. Of course, this means that 3 is subtracted from the expression $z^2 + 2z$.

There is a `factor` *command in Maple.*

```
factor(");
```
The factors of the expression assigned to the " symbol are displayed.

Check this result using the `expand` *command.*

```
expand(");
```
Always check your results. You may have made a mistake in entering the expression. On rare occasions, Maple makes errors as most large and complex programs do. These rare errors do not lessen the usefulness of such programs. Develop the habit of checking your Maple computations in the same way you have developed habits and methods for checking your hand computations.

An expression can be assigned to a variable name.

```
p := x^2 + 2*x - 3;
```
Note that p is assigned the value $x^2 + 2x - 3$.

You can always check what value is assigned to a variable.

```
p;
```
The value displayed is the value of p.

You can factor the expression p.

```
factor(p);
```
You should check to see that the factors of p are the same as the factors of $x^2 + 2x - 3$ given earlier.

Has factoring p changed p?

```
p;
```
The variable p has remained the same.

Solve the equation $x^2 + 2x - 3 = 0$ or $p = 0$.

```
solve(p = 0);
```
The solution is a set of two integers. These integers, as you know, are the solutions to the two equations $(x + 3) = 0$ and $(x - 1) = 0$.

You can plot or graph an expression. You can indicate the domain of the variable x in the expression p.

```
plot(p, x = -4..4);
```
Notice how the domain is specified. You *must* use the two periods rather than a comma. The graph will be displayed after a short time in a separate window.

Does the graph intersect the x-axis at the same values you obtained solving the equation $p = 0$? On the IBM under Windows and the Sun, you can check this by moving the cursor to the left point of intersection on the x-axis and clicking the mouse button. The horizontal coordinate displayed in the Plot window should be close to -3. Click on the Close box in the upper left corner of the Plot window to close it (press Esc under DOS). Do not save the Plot window if requested. You are again ready to enter Maple commands.

You can, if you wish, specify the range values.

```
plot(p, x = -4..4, y = -10..10);
```
Be sure there are two periods in the range specification. Notice that the tick marks on both the $x-$ and $y-$axes are integers. Click on the Close box to close the plot window or press Esc.

There is a common mistake made in entering the plot *command.*

```
plot(p, x = -2,3);
```
The error message clearly indicates where the problem occurs. When you see this message, you should first check to be sure you have entered the domain and range using periods.

You can easily change the domain of your graph.

```
plot(p, x = -2..3, y = -10..10);
```
Changing the domain and range of your graphs allows you to explore the behavior of an expression. You can move the Plot window by clicking and dragging its Title bar.

Editing on the Macintosh

You can use the powerful editing features of the Macintosh to speed your work with Maple. (Verbal directions on the right will appear in *italics*.)

Copy the last plot *statement.*

Use the mouse to move the cursor to the beginning of the previous **plot** *statement. Hold down the mouse button, drag the cursor to the end of the statement, and release the mouse button.*

You will have performed this action correctly if the entire **plot** statement is highlighted.

Now pull down the Edit menu by pointing at it and holding down the mouse button. Then select Copy by continuing to hold down the mouse button and dragging it down until Copy is highlighted. Release the mouse button.

An image of the **plot** statement is now saved.

You can paste this copied statement anywhere you wish in the Maple environment.

Move the cursor next to the last prompt and click the mouse button. Pull down the Edit menu and select Paste.

The **plot** statement should appear at the cursor position.

Edit this statement so that the domain of the plot is from −3 *to* 3.

To do this, move the blinking cursor to the right of −2 *in the* **plot** *statement. Now press the Backspace key to erase the 2. Type 3 and press Enter.*

Notice that you did not have to be at the end of the line for Maple to accept the entire statement when you pressed Enter.

If you pull down the Edit menu, you can see that there are special key combinations that can be used to accomplish copying and pasting. For example, if you highlight a word, you can copy it by holding down the key with the ⌘ on it and pressing C. You can speed your work by using these key combinations. The Undo command in the Edit menu is very useful if you are surprised by an edit operation. You can use the keys with arrows on them on your keyboard to move the cursor around the screen.

Editing on the IBM

Editing under Windows is essentially the same as for the Macintosh covered above. If you are working under DOS, you can display any statement you input by pressing the up-arrow key until the desired statement appears on the screen. You can then edit that statement using the left and right arrow keys along with the Insert and Delete keys. Once you have made the changes, you can execute the statement by pressing Enter. You need not be at the end of the line.

Ending a session.

```
quit;
```

You should end a session in the normal way. You can also end a session by selecting **quit** on the File menu or pressing F3 under DOS.

Extensions

Help Facility

The Help facility in the Maple environment is extremely useful. You already know how to access it using the **?** symbol.

The Macintosh version of Maple V allows you to access the Help facility by selecting Maple Help under the Windows menu at the top of the screen. You should make a point of learning to get around in the help facility and to consider using it anytime you have a question regarding commands, packages, etc. For example, if you click on the Help menu, you will access a help screen. Topics you can get help on are in the left column. When you click on one of these topics, a list of subtopics are displayed. Click on a subtopic in the second column and a list of topics that come under that category are displayed. If you click on any of these subtopics, a synopsis of that topic is displayed near the bottom of the screen. A full explanation of the subtopic can be accessed by clicking on the Help button at the bottom of the screen. You should experiment with this facility. Remember to click on the Go Away or Close box when you have finished with the window.

The IBM help facility under Windows is very similar to the Macintosh in many ways. You will notice that a Help menu is one of the main menus at the top of the screen. You can select Browser or Interface Help under this menu. The Browser facility is just like the help facility in the Macintosh environment. The Interface Help facility allows you to get information on a number of topics that are available under Windows. You obtain the information by clicking on any of the major headings displayed. In several of the topics listed, there is information given as well as other headings that contain further information.

The IBM help facility under DOS is accessed by pressing the F1 function key. You obtain the equivalent of the Browser selection under Windows. You select a topic or subtopic using the up and down arrow keys. The right and left arrow keys move you from column to column displaying subtopics. When you arrive at a column that does not contain an ellipses in the synopsis, you can press Enter to obtain a full explanation of that subtopic. As usual, the Esc key backs you out to the session window.

Macintosh Environment

The Main menu items in the Macintosh environment of Maple V have a number of helpful features. The Settings menu allows you to determine, among other things, how your output will be displayed. You use the Command menu to manipulate the most recently computed result. For example, if you enter

```
p := x^2 - 4;
```

and then select **Factor** from the Command menu, Maple will factor the expression.

You can use the **Save** and **Print** items of the **File** menu to save or print your session.

IBM Environment Under Windows

The pull-down menus under Windows and selection of

subitems are accomplished in the same way as on the Macintosh. For example, if you select Math Style under the Format menu, you can choose the size that Maple will use to display mathematics. Other items available are easily investigated by pulling down a menu and trying the items listed.

IBM Environment Under DOS

The F10 function key displays six menu items at the top of the screen. The subitems under these menus can be useful for a number of tasks and activities you wish to perform. You display the subitems by pressing Enter. You select a subitem by either highlighting it using the up and downarrow keys and pressing Enter or pressing the highlighted letter in a subitem. You move from menu to menu with the right and leftarrow keys. You can load, save, or print a session by selecting the appropriate subitem under the Session menu. Pressing Esc one or more times backs you out to the session window.

Additional Activities

Entering Expressions

Write each of the following expressions as you would enter them in Maple:

1. $\dfrac{1}{x-2}$

2. $\dfrac{1}{x} + \dfrac{5}{3x}$

3. $\dfrac{x-2}{x^4 - 3x^3 + 1}$

4. $\dfrac{x-4}{(x^2 - 2x - 7)^5}$

5. 2^x

6. 2^{x+5}

Exploring Expressions

Explore the following expressions using the **solve**, **factor**, and **plot** commands.

7. $x^2 - 5x + 6$

8. $x^2 - 4x - 12$

9. $6x^2 + x - 15$

10. $40x^2 - 131x + 84$

11. $90x^2 - 249x + 168$

2 Precalculus Algebra

This chapter presents many of the commands available in the Maple environment that are useful in a typical precalculus course. The major topic in such a course is the study of functions: their definitions, domains and ranges, asymptotes, graphs, and behaviors. Maple is an extremely powerful tool in this study and in other precalculus topics as well.

2.1 Solving Equations

Polynomial Equations

You've seen how to solve some equations. Let's look at some more.

```
q := 3*x^2 - 5*x + 2;
solve(q = 0);
```
Here you are entering two statements. Press Enter after each statement. This is $3x^2 - 5x + 2 = 0$. Use your factoring ability to check the solutions displayed.

Use subs *to check your result.*

```
subs(x=1, q);
```
The displayed result of 0 indicates that the value of q when $x = 1$ is 0. In a similar manner, you can check that $x = \frac{2}{3}$ is a correct solution also.

Sometimes solutions are irrational numbers.

```
solve(x^2 - 3 = 0);
```
Maple uses the radical symbol for square roots only (except under DOS). Fractional exponents are used for all others.

You can solve equations whose solutions are more complicated expressions.

```
solve(3*x^2 - 5*x + 1 = 0);
```
The two displayed solutions are, as before, separated by a comma.

You may wish to have decimal approximations for these solutions.

```
fsolve(3*x^2 - 5*x + 1 = 0);
```
The number of digits in the decimal approximations will depend upon the value of *Digits*.

  You can solve the equation $3x^2 + 5x + 1 = 0$ in a way that takes longer but provides the details of the steps required. The Maple commands needed in the step-by-step method are located in the student package.

You access the student package using the with *statement.*

```
with(student);
```
The commands available in this package are displayed.

Reenter the equation to be solved.

```
3*x^2 - 5*x + 1 = 0;
```
The equation is displayed in standard mathematical notation.

Now complete the square in x.

```
completesquare(", x);
```
The **completesquare** command is **cmpltsq** on older versions of Maple. In the equation returned, the lefthand side is displayed with the square completed.

You can isolate various parts of an equation.

```
isolate(", x - 5/6);
```
Notice that $x - \frac{5}{6}$ is isolated on the left side of the equation.

Now to the final solution.

```
isolate(", x);
```
This solution is the same as one of the solutions you obtained with **solve**. The step-by-step process has many educational advantages but also requires some thought on your part to supply missing details, if any. In this case, the **isolate** command obviously does not give the plus and minus values when taking the square root.

Some equations have complex solutions.

```
solve(x^2 + 1 = 0);
```
The complex number i may be represented as I. This can be difficult to read on the screen. You may wish to use the Font menu to change the font so that i is easier to discern.

You can solve equations whose solutions are more involved complex expressions.

```
solve(3*x^2 - 5*x + 7 = 0);
```

You need to be careful to look for i's in solutions since they can look like absolute value signs with certain fonts.

Again, you may wish to display the solutions in decimal form.

```
fsolve(3*x^2 - 5*x + 7 = 0);
```

The **fsolve** command finds only real number solutions.

The exact solutions can be approximated by decimals.

```
solve(3*x^2 - 5*x + 7 = 0);
```

This displays the two exact complex solutions as before.

You can assign this sequence of solutions to a variable.

```
s := ";
```

Recall the " mark represents the most recent computed result.

The variable s is a sequence with two elements. You can access each solution separately.

```
evalf(s[1]);
```

Notice the brackets around 1. The **evalf** command returns the floating-point approximation of the first solution in the sequence.

You can access the second solution as well.

```
evalf(s[2]);
```

The variable **s** contains the sequence of solutions. The second element of this sequence is designated as **s**[2] in Maple. Brackets (rather than parentheses) are required.

You can solve polynomial equations of higher degree than 2. You begin by assigning the polynomial expression to a variable.

```
q := 6*x^4 - 35*x^3 + 22*x^2 + 17*x - 10;
```

This is $6x^4 - 35x^3 + 22x^2 + 17x - 10$.

You can now solve the equation $q = 0$.

```
solve(q = 0);
```

You know that a polynomial of degree 4 will have at most four solutions. This equation has four rational solutions.

A slightly different equation ($q = 1$) can give a much different result.

```
solve(q = 1);
```

The solutions are clearly very complicated. You can see that portions of the displayed solutions involve numbers with percent (%) signs attached, like %6. Maple uses such percented numbers to represent expressions when displaying complicated solutions like these. The values of the percented expressions appear after the solutions.

You can get a much less complicated approxima-tion to the solutions us-ing the fsolve *command.*

```
fsolve(q = 1);
```
Notice that the solutions are between -1 and 6 and are real numbers.

You can use plot *to graph the expression* $q - 1$ *to resolve these conflicting results.*

```
plot(q - 1, x = -1..2);
```
The places where this graph crosses the x-axis are the so-lutions of the equation $q - 1 = 0$. The graph indicates that there are four real solutions that seem to correspond to the solutions obtained using **fsolve**. You can adjust the graphing window in the **plot** command to investigate this expression further. Here again, you have used the power of Maple to check the results you obtained with Maple. Click on the Close box before entering more Maple statements.

Other Types of Equations

You can solve many trigono-metric equations.

```
p := cos(x) - sin(x);
```
Notice that parentheses are required in using the trigono-metric functions.

Now you can solve the equa-tion $p = 0$, *which is equiv-alent to* $\cos(x) - \sin(x) = 0$.

```
solve(p = 0);
```
You may recall that trigonometric equations often have an infinite number of solutions. The solution displayed on the screen, **1/4 Pi** or $\frac{\pi}{4}$, may not be the only solution. The **solve** command usually returns a single solution to nonpolynomial equations even if there are many solutions.

Graphing the expression can assist you in determin-ing the complete solution set.

```
plot(p, x = 0..2*Pi);
```
The symbol π is represented by Pi with a capital P. Notice that there is more than one solution.

You can use the fsolve *command to find other solutions.*

```
fsolve(p = 0, x, 1.5..4);
```
Notice the comma after **x**. The 1.5 . . 4 indicates which interval **fsolve** searches for a solution. The number dis-played should be a multiple of π.

What multiple of π is it?

`evalf("/Pi);`
Thus, the solution is approximately $\frac{5\pi}{4}$, which is π units from the previous solution. Can you write down the solution set in set notation based on this information?

You can solve logarithmic equations also.

`solve(ln(x) + ln(x+1) = ln(2));`
$\ln(x)$ is the natural logarithm of x. Thus, $\ln(2)$ is $\log_e(2)$ and is approximately 0.693172. The solutions to the equation are given as 1 and -2. The number -2 cannot be a solution, since $\ln(-2)$ is undefined. You may wish to solve this equation using paper and pencil methods to see why -2 arises as a solution. Logarithms to bases other than e can be entered using the standard conversion formula. For example,

$$\log_{10}(x) = \frac{\ln(x)}{\ln(10)}$$

Exponential equations are easy to enter and solve.

`solve(2^x = 5);`
The exact solution is displayed using natural logarithms. You should use the **evalf** command to obtain a decimal approximation to the solution.

Use `fsolve` *to find an approximate solution to this equation.*

`fsolve(2^x = 5);`
Notice that the solution is different from the solution you obtained using **evalf**. Maple uses two different methods to obtain this approximate result. The results differ in the tenth place. Better approximations can be obtained by setting *Digits* to a larger integer value.

Inequalities

You can solve inequalities.

`solve(x^2 - 5*x < 0);`
The solution set for this inequality is the interval $(0, 5)$, or the set $\{\mathbf{x}|\ \mathbf{0<x<5}\}$. The solution is displayed by Maple as $\{\mathbf{x<5},\ \mathbf{0<x}\}$, a form of set notation meant to represent set intersection.

Maple also uses a form of set notation meant to represent set union.

`solve(x^2 - 5*x >= 0);`
Notice how the $\geq$ symbol is represented in Maple. Maple

displays {**x<=0**}, {**5<=x**}. The solution uses two sets of braces to indicate the union of two sets as compared with the intersection notation which uses only one set of braces. You should be careful to look for these differences in notation when solving inequalities.

You can also solve systems of equations.

```
s := solve({x+y=5,x-y=2}, {x,y});
```
Here, Maple solves a set of equations (in the first pair of braces),

$$x + y = 5$$
$$x - y = 2$$

for a set of variables (in the second set of braces). The solution set is displayed with an x value and a y value.

Use subs *to check your results.*

```
subs(", x+y=5);
```
Here, the quoted expression

$$\{y = 3/2, x = 7/2\}$$

gives the values of x and y substituted in the equation **x+y = 5**. This expression is equivalent to

subs(x=7/2, y=3/2, x+y=5)

but is quicker to enter.

You can check the solution of the system of equations graphically.

```
plot({5-x, x-2},x=-4..4);
```
Notice that you must write the equation $x + y = 5$ in the form $y = 5 - x$ and $x - y = 2$ in the form $y = x - 2$ to plot the graphs. Also, the braces are required when you plot a set of more than one function on the same axes.

A slight change in the equations gives a different result.

```
solve({x+y=5.,x-y=2}, {x,y});
```
Notice the decimal point after the 5 in the first pair of braces. Maple uses a numerical method that yields approximate solutions if one of the numbers in the system of equations is in floating-point format. In this case, the approximations have the same value as the exact rational solutions displayed in the previous item.

Additional Activities

Explore the following polynomials using the **factor**, **solve**, **fsolve**, **evalf**, and **plot** commands.

1. $x^4 - x^3 - 5x^2 + 12$
2. $2x^3 - 13x^2 - 4x + 60$
3. $8x^2 + 2x^3 - x^4$
4. $2x^4 - 5x^3 + 10x - 12$
5. $x^5 - x^4 - 15x^3 + x^2 + 38x + 24$
6. $x^5 - x^4 - 15x^3 + x^2 + 38x + 10$

2.2 Rational Expressions

You can assign a rational expression to a variable.

```
r := 1/(x + 1) - 1/(x - 1);
```
Since you cannot enter horizontal fraction bars, you need to be careful to delineate the numerators and denominators clearly using parentheses. Fortunately, Maple displays this rational expression in standard two-dimensional format.

You can simplify this expression.

```
simplify(r);
```
Maple adds the two rational expressions and displays the result in two-dimensional form.

Extract the denominator from this expression using the denom *command.*

```
denom(simplify(r));
```
This is one of the many places you should consider using the copy feature or recalling previous lines to reduce the amount of typing. The denominator is displayed in factored form as it was displayed in the last item. You can use the **expand** command to display this denominator as a polynomial in standard form.

The numerator can be extracted with the numer *command.*

```
numer(simplify(r));
```
The number -2 is the numerator of this rational expression.

Graph this expression using the `plot` *command.*

`plot(r, x=-10..10);`
You know that this expression should have asymptotes from your experience in precalculus mathematics. However, the steep lines that appear to be asymptotes near -1 and 1 are only lines joining points on the graph.

You can force Maple to plot just the points that it computes for a graph of an expression.

`plot(r, x=-10..10, style = POINT);`
The **POINT** designation must be in capital letters. The graph is now displayed without connecting the computed points. There are no apparent asymptotic lines on the displayed graph.

Use the `solve` *command to determine where the asymptotes should be on the graph.*

`solve(denom(r));`
The two solutions give the location of the asymptotes. They can be checked against the graph of the expression.

Graph the expression with a smaller domain that still includes the asymptotes.

`plot(r,x=-2..2,y=-30..30,style=POINT);`
Maple now displays a graph that more closely represents the known features of the graph of the expression.

Let's look at another rational expression.

`s:=(x^2 + 5*x + 6)/(x^3 + 2*x^2 - x - 2);`
Notice the parentheses around the numerator and denominator of the expression

$$\frac{x^2 + 5x + 6}{x^3 + 2x^2 - x - 2}$$

You can factor both the numerator and denominator of this rational expression.

`factor(numer(s));`
`factor(denom(s));`
Here, you are entering two statements (pressing Enter after each statement). You can see that the numerator and denominator can be factored.

Simplify the expression s.

`simplify(s);`
Now the original expression s is seen to be in lowest terms. You should be aware that this simplification is correct only if $x \neq -2$.

You can graph this expression.

`plot(s, x=-3..3, y=-30..30);`
Notice that there is no vertical asymptote at $x = -2$, even

though you saw that $x + 2$ was a factor of the denominator. There is no vertical asymptote at $x = -2$ because this factor also appears in the numerator. The Maple graph seems to be defined at $x = -2$, which is incorrect since the denominator is zero at $x = -2$. Maple simplifies an expression before graphing it. This can result in the loss of information. In this case, the graph will never show that the function is undefined at $x = -2$. You should continue to be aware of possible discrepancies between Maple graphs (which are just connected points) and actual graphs.

Additional Activities

Determine the zeros and asymptotes of the following rational expressions using the **factor** and **solve** commands.

1. $\dfrac{2x - 3}{x^2 - 9}$

2. $\dfrac{2x + 3}{x - 1}$

3. $\dfrac{x^2 - 2x - 8}{x^2 - 2x}$

4. $\dfrac{x^2 + 3x - 10}{4x + 20}$

5. $\dfrac{x^2 - 1}{x + 2}$

6. $\dfrac{x^2 - 1}{x^3 - 1}$

7. $\dfrac{2x^2 - 3x - 2}{x^2 - 5x}$

8. $$\dfrac{x^2 - 2x + 1}{x^4 - 1}$$

9. $$\dfrac{4x^3 - 5x^2 + 3x - 6}{2x^2 + 3x + 5}$$

10. $$\dfrac{2x^3 - 7x^2 + 7x - 2}{2x^2 + 5x - 3}$$

2.3 Defining Functions and Procedures

Functions of One-Variable

You can use the Maple procedure facility to create function definitions. A procedure performs a task that is described in a set of instructions. A function is a procedure that returns a value specified by its set of instructions. You can use this facility to define functions such as:

$$f(x) = x^2 + 3x - 5$$

$$g(x) = \cos(x) - x \ln(x)$$

$$h(x) = \dfrac{\cos(x)}{x^2 + 3x - 21}$$

You can define the real-valued function f whose rule is $f(x) = x^2$.

```
f := x -> x^2;
```
You enter the mapping arrow (->) by typing a minus sign (-) followed by a greater than symbol (>) with no space between them. The function is named f and maps or takes x to x^2. Thus, x^2 is the rule for the function. This procedure has one variable x and one task (to square x). It is a function because it returns exactly one value. Maple echoes the procedure definition when you enter this line.

You can obtain function values for f.

```
f(2);
```
This statement asks Maple to evaluate and display the function value of f at $x = 2$.

You can use variable names in a function evaluation.

```
f(a+b);
```
The value displayed is the value of f at $x = (a + b)$.

You can use the **expand** command to display this function value without parentheses.

You can also use defined functions in algebraic expressions such as
$$\frac{f(x+h) - f(x)}{h}.$$

```
(f(x + h) - f(x))/h;
```
Parentheses must be placed around the entire numerator, since you are entering this expression on one line.

The value displayed can be simplified.

```
simplify(");
```
Maple removes the parentheses, combines like terms, and reduces fractions to lowest terms. Notice that the simplified expression is equivalent to the original expression as long as $h \neq 0$.

Create a procedure using proc *to define more complicated real-valued functions such as*
$$f(x) = \begin{cases} x^2 & \text{if } x > 3 \\ x - 5 & \text{if } x \leq 3 \end{cases}.$$

```
f := proc(x) if x > 3 then x^2 else
     x - 5 fi end;
```
This single statement cannot fit on one line in this book. This is indicated by the indentation of the second line. However, you should type it on a single line. In this procedure, if the domain value x is greater than 3, then $f(x)$ is determined by the rule x^2. On the other hand (**else**), if x is less than or equal to 3, then $f(x)$ is determined by the rule $x - 5$. The **if** phrase is terminated by **fi** (**if** spelled backwards). You must enter this statement carefully.

You can obtain function values as before.

```
f(2);
f(5);
```
Remember to press Enter after each semicolon. You should check to make sure the function is using the appropriate rule in each case.

You can use plot *to graph this function.*

```
plot(f);
```
Does the graph of the function clearly show the two pieces of the graph? You should graph this function again using the **style=POINT** feature in **plot** to see a more accurate representation of the function. You may have to select suitable domains to see the graph clearly in some complicated functions.

Unassign the definition of f.	`f := 'f';` This is similar to unassigning the value of a variable.

Functions of Several Variables

Functions of several variables occur in the latter part of the calculus sequence. They are written in mathematical notation much like one-variable functions:

$$f(x, y) = x^2 + y^2 - 3$$

is one example.

Define this function in Maple.

`f := (x,y) -> x^2 + y^2 - 3;`
Here, each domain value of the function is an ordered pair of numbers (in parentheses) and the rule on the right of the mapping arrow uses two variables.

You can find values for such functions.

`f(2,5);`
The function value for the pair of numbers (2, 5) is $2^2 + 5^2 - 3$. The number displayed is 26.

Such functions can be plotted in three-dimensions.

`plot3d(f, -2..2, -2..2);`
This three-dimensional plot uses values for x and y in the xy-plane from -2 to 2 in both the x and y directions. Additional information about three-dimensional graphing is covered at the end of Chapter 3.

Automating Commands in Maple

Frequently, you will want to repeat some particular command or set of commands several times with slightly different values. You do this with a looping structure using the **for** and **do** commands.

Print a table of values.

`for k from 1 to 3 do print(k, k^2); od;`
This prints out three pairs of numbers. Notice that the index moves from 1 to 3 by steps of 1. The command that is repeatedly performed is a **print** command and it is enclosed in the pair **do/od** just as **if/fi** begin and end a conditional command.

Recall the reverse roles of the Enter and Return keys on the Sun.

Retype this line in more readable multi-line form. Be sure to use the Return key on the Macintosh or the combination of the Shift-Enter keys under Windows at the end of the first three lines. Press Enter after the last line.

```
for k from 1 to 3
  do
    print(k, k^2);
  od;
```

(On the Sun, you **must** highlight all of these lines before pressing Return.) The indentation in these lines is a structured format commonly used in programming. When you press Enter, all the lines will be performed, causing three pairs of numbers to be displayed on the screen.

Modify this looping structure to step by one-half.

```
for k from 1 by .5 to 3
  do
    print(k, k^2);
  od;
```

Again, remember to press Return (Shift-Enter under Windows) at all but the last line. Press Enter at the last line to define this procedure. Notice that there are now five pairs of values since the step size is one-half. You may wish to use the editing capabilities of your computer rather than retyping the entire program.

An Automating Procedure

You can use this looping structure as part of a procedure. Such procedures are often called programs. See Appendix A for an alternate way of programming under DOS.

Define maketable.

```
maketable :=
  proc(n)
  for k from 1 by .5 to n
    do
      print(k, k^2);
    od;
  end;
```

Remember to press Return at the end of every line but the last on the Macintosh (Shift-Enter under Windows) as you enter this procedure. The procedure **maketable** uses **n** as its argument. Press Enter to define this procedure. Notice

that only a compact version of the procedure definition is
displayed.

Let's use this procedure.

```
maketable(5);
```
As you might expect, a table of squares is displayed.

*Generalizing the
maketable procedure.*

*Change the second and third lines of the procedure defini-
tion as follows.*

```
maketable :=
  proc (n, increment)
  for k from 1 by increment to n
    do
      print(k, k^2);
    od;
  end;
```

After you have made the changes, press Enter to define the
procedure. Again, a compact version of the procedure is
displayed.

*Let's use this new proce-
dure.*

```
maketable(3, .2);
```
Are the values displayed what you would expect? The step
size is now .2 and the last pair is 3, 9.

Automated Plotting

You can define a procedure to plot several related functions
on the same axis.

*Graph a set of polynomi-
als.*

```
polyplot :=
  proc (n)
  polys := NULL;
  for k from 1 to n
    do
      polys := polys, x^k;
    od;
  plot({polys},x=-10..10,y =-10..10);
  end;
```

Press Enter to define the procedure. This procedure creates
the polynomials in the **for/do** loop. The **plot** command

displays this set of polynomials ({**polys**}) on the same axes.

Let's use this procedure.   `polyplot(3);`

Was the displayed plot what you expected? The straight line, quadratic, and cubic functions are graphed together.

You can generalize this procedure with additional arguments in many ways. For example, the domain and range of the Plot window can appear as arguments.

Extensions

Automating Commands

The **for** and **do** commands allow you to repeat a process as many times as you need. For example, you could use the following commands to display the interest and amount paid toward the principal of a loan. The monthly payments on a loan of $8,000 at 12% are $210.67 for 48 months.

```
prin := 8000;
months := 48;
for k from 1 to months
  do
    intr := prin * (12/1200);
    pay := 210.67 - intr;
    prin := prin - pay;
    print(intr, pay);
  od:
print('Principal remaining is', prin);
```

The colon after **od** is important if you wish to display only the interest and amount paid toward the interest. Otherwise, a semicolon will cause each of the assignment statements to be displayed everytime through the loop. It is also important to note that backquotes (') are needed around a string that is to be displayed as in the last **print** command. To execute this set of commands, you would press Enter with the cursor somewhere in the program. If you are using a Sun, you **must** highlight a set of commands and press Return to perform all the highlighted commands in order. Under Dos, the program would execute after the line **od:** is entered. You would have to enter the final line to complete the task. Entering, defining, and executing programs varies from computer to computer. You should check the Maple V manuals for your computer for details.

It is not difficult to generalize this routine into a program that would allow you to input the principal, rate, number of monthly payments, and the amount of each monthly payment. You can then display the monthly interest and amount paid toward the principal for the number of months you choose. Finally, you can display the principal remaining after making this number of monthly payments.

```
payment :=
  proc(p, rate, m, mopay)
  prin := p;
  months := m;
  for k from 1 to months
    do
      intr := prin * (rate/1200);
      pay := mopay - intr:
      prin := prin - pay;
      print(intr, pay);
    od:
  print(`Principal remaining is`,prin);
  end;
```

Once you enter this program, you can run it with various values as you wish. For example, to produce the same results as in the previous program, you would enter

```
payment(8000, 12, 48, 210.67);
```

Additional Activities

1. Enter the function $f(x) = \dfrac{x^2}{x^2 + 2}$ and graph it.

2. Enter the function of two variables $f(x, y) = x^2 + y^2$ and determine the value of the function at the points $(3, -2)$ and $(\sin(2), \cos(2))$.

3. Enter the function $f(x) = \text{sign}[\cos(x)]$. Graph this function. What values does this function take on most frequently?

4. Enter the function $f(x) = \dfrac{x^2 - 3x + 5}{\sin(x) + 1}$. Use Maple to determine whether this function is defined at $\dfrac{-11\pi}{2}$.

5. Develop and enter a function for the length of the line segment between two points (a, b) and (c, d). Use this function to determine the length of the line segment between $(2, 5)$ and $(-3, 7)$ and the length of the line segment between an arbitrary point (r, s) and $(2, 3)$. (Be sure to deassign all variables you use.)

2.4 Additional Features of the `plot` Command

The **plot** command has several options that allow you to create many types of graphs.

You may recall that Maple allows two or more functions to be graphed on the same axes.

```
plot({x^2,2*x + 5},x=-10..10);
```
You should watch carefully as the functions are drawn so that you can tell which graph is associated with x^2 and which is associated with $2x + 5$. In this case, you can easily tell the parabola from the straight line. You can change the Plot window if you wish to examine parts of these graphs (the intersections) more carefully by selecting suitable values of x. Remember, Esc gets you out of the plot window.

Parametric Form

You can graph curves expressed in parametric form. For example, the two equations

$$x(t) = t - 1 \text{ and } y(t) = t^2$$

give the x and y coordinates of a parabolic curve based on a parameter (dummy variable) t. If you solve these two functional equations for y in terms of x by eliminating the parameter t, you will find that $y = (x + 1)^2$, the equation of a parabola.

Use the plot *command with special grouping symbols, [ ], to graph the parametric curve just defined.*

```
plot([t - 1, t^2, t=-2..2]);
```
Notice the placement of the brackets ([ ]) around the two parametric function rules and the specification of the domain of t.

Adjust the graphics window using the standard method for specifying the horizontal and vertical axes.

```
plot([t - 1,t^2,t=-2..2],-5..5,-2..10);
```
The curve will still appear between -3 and 1 on the horizontal axis, since the parametric rule $t - 1$ (the first coordinate) has domain $[-2, 2]$. The axes, however, have values between $[-5, 5]$ horizontally and $[-2, 10]$ vertically. The function rules and the domain for the parameter t appear together inside the brackets, followed by the horizontal and vertical specifications for the graphics window. The **x** and **y** are not placed on the axes, since they are omitted from the domain and range arguments.

You can graph more complicated curves using this parametric approach.

```
plot([t - sin(t),1 - cos(t),t=0..2*Pi]);
```
The parametric functions are $x(t) = t - \sin(t)$ and $y(t) = 1 - \cos(t)$. The domain of the parameter t is $[0, 2\pi]$. This curve is called a cycloid, which is easy to describe parametrically but is very complicated to describe in standard function form.

Polar Form

The parametric graphing feature allows you to graph functions in polar coordinates.

```
plot([sin(t),t, t=0..Pi],coords=polar);
```
This graphs the polar coordinate function $r = \sin(t)$. You may recall that this function describes a circle. The graph, however, is oval and does not look like a circle.

You must use a graphics window with the appropriate aspect ratio.

```
plot([sin(t), t, t=0..Pi], -3..3, -2..2,
coords=polar);
```
The graphics window varies in size from computer to computer. One size is 6 units wide and 4 units high. This gives an aspect ratio of 3 to 2, or 1.5. That is the choice in the plot statement above. If the graph displayed does not look like a circle, you need to determine the aspect ratio of your computer and use those numbers so that the figure looks like a circle.

Some curves are more easily described using polar coordinates.

```
plot([1 + cos(t), t, t=0..2*Pi], -3..3,
-2..2, coords=polar);
```
This graph is called a cardioid, since it looks like a heart. It is the graph of $r = 1 + \cos(t)$. If you are using DOS, go on to the next section.

Except in DOS, you can adjust the size of the Plot window once the graph is displayed.

Place the arrow cursor in the lower right corner of the Plot window until the cursor becomes a double arrow. Then hold down the mouse button (left two buttons on the IBM under Windows) and drag it toward the middle of the screen. You should shrink the size of the Plot window until it takes up about one-half the screen. Be sure not to click on the Close box in this case. Rather, click on the Session window (the region under the graphics region) to return to the standard Maple environment.

Let's graph another function and display it side by side with the previous function graph.

```
plot([1 + sin(t), t, t=0..2*Pi], -3..3,
-2..2, coords=polar);
```
Shrink this Plot window to about half the size of the screen. Now use the Title bar to move this Plot window to the right half of the screen. You can use the Title bars on these two graphs to position the them as you wish. Often,

comparisons of similar graphs can give you insights into problems.

2.5 Putting It All Together

In this section, you will use combinations of Maple commands you have learned to investigate polynomial and rational functions.

A Rational Function with Asymptotes

The graphs of rational functions can have interesting features such as horizontal, vertical, and oblique asymptotes. You can investigate these asymptotes by looking at the zeros of the denominators and at the behavior of the functions as the variable approaches $\pm\infty$. You will investigate the following rational function

$$f(x) = \frac{3x^3 - x^2 - 3x + 5}{x^2 - 2x - 1}$$

First you define the function.

```
f:=x -> (3*x^3-x^2-3*x+5)/(x^2-2*x-1);
```
Notice the parentheses around both the numerator and the denominator.

You can graph the function to obtain an overview of the behavior of the function.

```
plot(f);
```
Looking at the graph, there appear to be one zero and two vertical asymptotes.

Restrict the domain to obtain more detailed information about the graph.

```
plot(f, -2..3);
```
This graph still doesn't seem to reveal enough information.

You can change the range as well.

```
plot(f, -2..3, -50..50);
```
The behavior of the function between the asymptotes is now more apparent. The seemingly vertical lines are, as you recall, calculated points that are connected. They are not truly asymptotes. You may wish to check this by using the **style = POINT** feature.

Determine the exact location of the vertical asymptotes.	`solve(denom(f(x)) = 0);` The vertical asymptotes can occur only at the zeros of the denominator. The values displayed are exact, but how would you graph them on the x-axis?
You can find floating-point approximations for these two values.	`fsolve(denom(f(x)) = 0);` Although exact values are useful at times, approximate values are also useful.
You may have noticed that the graph of the function seemed to be a straight line away from the asymptotes.	`quo(numer(f(x)), denom(f(x)), x);` Check to be sure there are sufficient parentheses that match. The **quo** command returns the polynomial part of the quotient. As you can see, this partial quotient is a linear rule.
You can check that this is an oblique asymptote.	`plot({3*x + 5,f(x)},x=-5..5,y=-50..50);` Remember, plots can take a while depending on your computer. Is the graph of the function close to the line away from the asymptotes? Notice that the function graph intersects the oblique asymptote. You may wish to examine the graph more closely to the left of -5 by further adjusting the x and y values. You might try $x = -20..0, y = -20..10$.
You can also locate the x- and y-intercepts.	`solve(f(x) = 0);` Once again, it is hard to tell what these possible x-intercept values are, but, if you look carefully, two of them are complex.
You can find an approximation of the real root.	`fsolve(f(x) = 0);` Notice that this point is to the left of both asymptotes.
The y-intercept is easy to determine.	`f(0);` The value of the y-intercept occurs when $x = 0$. You can easily make this computation in your head in this case.
Locate the intersection of the graph of the function and the oblique asymptote.	`fsolve(3*x + 5 = f(x));` We laughed, too. Do you believe this answer? Can you check this answer using pencil and paper?

Finding the Roots of Polynomials

You can use the graphing, factoring, and equation-solving capabilities of Maple to investigate the roots of polynomials.

Let's look at a fifth degree polynomial.

```
p := x -> 12*x^5 + 32*x^4 - 57*x^3
     - 213*x^2 - 104*x + 60;
```
This is $12x^5 + 32x^4 - 57x^3 - 213x^2 - 104x + 60$. Again, notice the use of the asterisk to indicate multiplication between coefficients and variables.

You can display this polynomial function.

```
p(x);
```
Maple displays the function in standard mathematical notation.

Graph this polynomial.

```
plot(p);
```
This graph gives few details of the behavior of the polynomial between -5 and 5. The graph of this polynomial, however, does appear to cross the x-axis between -5 and 5.

Focus your attention on a smaller interval.

```
plot(p, -5..5);
```
A few more details are discernable from this new graph. It shows that the polynomial may be zero at several points between -5 and 5. What is the largest value of a tick mark on the y-axis? You should think about restricting the y values so that more detail will appear.

You know how to specify the second coordinate values.

```
plot(p, -5..5, -10..10);
```
There appear to be two zeros to the right of the origin and at least one zero to the left. Further investigation will show that this plot does not give an accurate picture of the behavior of the function. Such plots can occur because Maple uses only 25 points to start its plot of an expression.

Try a different set of y values to investigate further the behavior of the polynomial.

```
plot(p, -5..5, -100..100);
```
This graph gives more information and seems to show the

complete behavior of the function between -5 and 5. How many zeros do there appear to be? What is the degree of the polynomial?

You can magnify the graph of the function to the left of the origin.

```
plot(p, -2.5..0, -10..10);
```
Here you can more clearly see that the graph just touches the x-axis at $x = -2$. You might reflect on what this means.

Look at the graph to the right of the origin.

```
plot(p, 0..3);
```
The location of the first of the two zeros to the right of the origin can more easily be approximated now.

Try to factor this polynomial.

```
factor(p(x));
```
Maple has factored this polynomial completely. How many linear factors are there? Can you determine the exact values of the rational zeros of the polynomial?

Use the solve *command to determine the zeros of the polynomial.*

```
solve(p(x) = 0);
```
Notice the zeros appear in the same order as the factors. How many times does -2 appear? How many times did the factor $(x + 2)$ appear?

You have used the **plot** command to investigate the behavior of the polynomial in a situation where all the zeros were rational.

Let's look at another polynomial.

```
q := x -> x^5 + 4*x^2 - 3*x + 5;
```
Be sure to enter the polynomial function carefully.

Graph the polynomial.

```
plot(q);
```
The graph seems to look the same as the first graph of p just given.

Graph the polynomial from -5 to 5.

```
plot(q, -5..5);
```
This time the polynomial seems to cross the x-axis once between -5 and 5.

You can get a better picture of the graph.

```
plot(q, -5..5, -100..100);
```
Now it is clear that the polynomial crosses the x-axis only once.

You can obtain a more detailed graph.

```
plot(q, -5..5, -10..10);
```
Much of the graph is cut off above the line $y = 10$. You can, however, more clearly see the behavior of the polynomial near its zero.

Let's factor the polynomial.

```
factor(q(x));
```
Surprise. Maple is unable to factor this particular polynomial. You know it must have at least one real zero from the Fundamental Theorem of Algebra. Thus, there should be at least one linear factor.

Hope is not lost. You can try the solve *command.*

```
solve(q(x)= 0);
```
The **RootOf** the same polynomial in z indicates that Maple could not find an exact solution.

Let's stay with it. You can try the fsolve *command.*

```
fsolve(q(x) = 0);
```
The approximate value is displayed. It is a negative value that matches the information on the graphs. Without further investigation, you might conclude that the other zeros are complex zeros. You might wish to examine the symbolic form of the polynomial to verify for yourself that there are no other real zeros outside the interval $[-5, 5]$.

The rational zeros of a polynomial can be found using the Maple **solve** or **factor** commands. Irrational and complex zeros are sometimes found by Maple. Polynomials of high degree (5, 8, 10) can be difficult to investigate. The graphing approach is a very useful tool in such situations.

Let's look at one last example.

Enter the polynomial.

```
r:=x -> 2*x^5+11*x^4+2*x^3-51*x^2-14*x+60;
```
This is $2x^5 + 11x^4 + 2x^3 - 51x^2 - 14x + 60$. Be careful as you enter this polynomial.

You can factor first.

```
factor(r(x));
```
Maple factors this polynomial into linear and quadratic factors with integer coefficients. Can you determine the nature of the zeros?

Use the `solve` *command.*   `solve(r(x)=0);`
All the zeros are displayed. How do they match up with the factors displayed before?

It might be interesting to   `plot(r, -5..5, -100..100);`
graph this polynomial.   Does this initial graph give you information that matches the displayed zeros? You may wish to plot this polynomial for various values of x and y.

The **factor** and **solve** commands may give you all the information you need. When they don't, then it is useful to investigate the polynomial using the **plot** command. You need to be flexible in your use of these tools.

Additional Activities

Graph the following functions.

1. $x = \sin(t), y = 2\cos(2t)$

2. $x = t^2, y = \cos(t)$

3. $x = t, y = t^2$

4. $r = 2\sin(3t)$

5. $r = 1 - 2\sin(t)$

6. $y = \cos(x)$

Graph the following functions indicating their zeros and their horizontal and vertical asymptotes, if any.

7. $f(x) = \dfrac{2x^4 - 2x^2 + x + 5}{x^2 - 3x - 5}$

8. $f(x) = \dfrac{2x^4 + 7x^3 + 7x^2 + 2x}{x^3 - x + 51}$

9. $g(x) = \dfrac{2x^3 + 3x^3 + x^2 + 2}{x^3 - x^2 + 21}$

10. $p(x) = x(x^2 - 3)(x^2 - 8)$

11. $r(x) = 999x^3 + 780x^2 - 5428x + 3696$

12. $g(x)\dfrac{x^5 - x^4}{x^5 - 6x^4 + 5x^3 + 26x^2 - 48x + 18}$

2.6 Matrices and Matrix Commands

The Maple operators you will need for this section are contained in the linear algebra package called **linalg**. You will need to load this package in order to study matrices and linear transformations and the solution of systems of linear equations.

You use the with *statement to access the* linalg *package.*

`with(linalg);`
The commands that are loaded into the Maple environment are displayed on the screen.

You enter a matrix with the matrix *command.*

`A:=matrix(4,4,[[1,2,3,4],[2,3,0,-5],`
`[2,-1,1,1],[-2,2,0,-5]]);`
The first two arguments of this command are the row and column dimensions of the array. The third argument of this command is the set of matrix entries. The bracketed sets of numbers represent the row entries of the matrix. Thus, the first row has four entries: $1, 2, 3, 4$. The fourth of the four rows has entries $-2, 2, 0, -5$. The bracketed row entries are nested inside brackets.

You can attempt to display the matrix.

`A;`
Notice that **A** is returned. The array data structure is not displayed, only its name.

You can, however, display the matrix entries.

`evalm(A);`
Notice that this command displays the matrix in standard row and column format.

Add two matrices.

`add(A, A);`
Here you are adding the matrix A to itself. Notice that each entry of the displayed matrix is twice the corresponding entry of the matrix A.

You can interchange two rows.	`B := swaprow(A, 3, 4);` You may wish to display the matrix *A* again to compare it with *B*. Notice that this new matrix is assigned to *B*. The **swapcol** command is used to interchange two columns.
You can add A and B.	`add(A, B);` The last two rows of this matrix are the same. You can obtain the same result using **evalm(A + B);** .
Save this matrix for later use.	`D := ";` *D* is a 4 × 4 matrix whose last two rows are the same.
You can add a multiple of one row to another.	`addrow(D, 1, 2, m);` Notice that row 2 is replaced by *m* times the elements of row 1 plus the elements in row 2.
Create a matrix that is a linear combination of A and B.	`add(A, B, 1, 2);` The entries of *A* are multiplied by 1 and added to 2 times the entries of *B*. You can use this extended form of the **add** command to subtract two matrices.
Find the inverse of A.	`inverse(A);` In DOS, Maple displays rational numbers on a single line with a slash if either the numerator or the denominator has one digit.
You can check the inverse computation.	`multiply(A, ");` As you would expect, a 4 × 4 identity matrix with 1s on the main diagonal is displayed.
Maple can transpose matrices.	`transpose(A);` You may wish to display *A* to check the result. Notice that the rows of *A* have become the columns of the transposed matrix.
The determinant of a matrix is easily obtained.	`det(A);` The determinant is a multiple of the denominators of the inverse matrix.

You may recall that the determinant of a matrix with two identical rows is zero.

```
evalm(D);
det(D);
```
As you might suspect, the determinant is 0. What should this imply for the inverse of *D*?

You can attempt to compute the inverse of D.

```
inverse(D);
```
An error message is displayed because the matrix is singular.

Additional Activities

1. Find the inverse, transpose, and determinant of the following matrix:

$$\begin{bmatrix} 3 & 2 & 4 \\ 4 & -2 & 6 \\ 8 & 3 & 5 \end{bmatrix}$$

2. Find the inverse, transpose, and determinant of the following matrix:

$$\begin{bmatrix} 7 & -8 & 1 & 2 \\ 21 & 4 & 3 & -1 \\ -35 & 8 & 3 & -2 \\ 14 & 16 & 0 & 1 \end{bmatrix}$$

3. Solve the following system of equations.

$$3w + x + 7y + 9z = 4$$
$$w + x + 4y + 4z = 7$$
$$w + 2y + 3z = 0$$
$$2w + x + 4y + 6z = -6$$

Ans. $w = 1$, $x = -6$, $y = 10$, $z = -7$

4. Let $A = \begin{bmatrix} 1 & 3 & -2 \\ -4 & 1 & 5 \\ 2 & 3 & -1 \end{bmatrix}$. Find the determinant and the determinant of the transpose of *A*.

5. Find the determinant of the matrix $A = \begin{bmatrix} 1 & 2 & 3 \\ 4 & 5 & 6 \\ 7 & 8 & 9 \end{bmatrix}$. Use the **addrow** command to change A by first replacing row two with -4 times the first row added to the second row and then replacing row three with -7 times row one added to the third row. Finally, replace row three with -2 times row three. Is the bottom row of the modified matrix all zeros? Does this fact verify the determinant value you obtained with the **det** command?

6. Find the transpose of both A and A^{-1} where

$$A = \begin{bmatrix} 3 & 2 & -5 \\ -1 & 4 & 2 \\ 2 & -3 & 1 \end{bmatrix}.$$

How do they compare?

3 Calculus

3.1 Calculus I

Maple contains a number of calculus commands or operators. These include differentiation, integration, and limit-taking commands.

Limits of Functions

You can take the limit of a function as the variable approaches a fixed number for a defined function f.

```
f := x -> (x^2 - 4) / (x - 2);
limit(f(x), x = 2);
```

Remember to press Enter after each semicolon. This is $\lim_{x \to 2} f(x)$, where $f(x)$ is defined to be

$$\frac{x^2 - 4}{x - 2}$$

Notice that $x = 2$ indicates the value that the variable x will approach. The displayed value is the limit. You can see that this function is undefined at $x = 2$ but the limit exists there.

Graph the function to check if this limit seems correct.

```
plot(f, -5..5);
```

Notice that the graph appears to be a straight line even though you know that the function is undefined at $x = 2$.

Can you draw an accurate graph of this function on the interval $[1, 3]$? Does the limit displayed before appear to be correct?

Factor the function as an additional check.

```
factor(f(x));
```

Can you explain why this factored form is not an accurate replacement of $f(x)$? Look at the definition of f and observe that it is undefined at $x = 2$. Is the factored representation undefined at $x = 2$?

You can find limits of more complicated functions.

```
f := x -> (x - 4)/(sqrt(x) - 2);
limit(f(x), x = 4);
```

The function

$$f(x) = \frac{x - 4}{\sqrt{x} - 2}$$

is not defined at $x = 4$. Does the limit displayed seem correct?

Graph the function to check the answer.

```
plot(f, 0..5);
```

Once again, the function graph is slightly flawed. Can you draw the correct graph of the function? Does the graph resemble a straight line or is it curved (except at $x = 4$)?

You must remember that the **plot** command in Maple computes only a finite number of points and then draws a smooth curve joining the points. Thus, it can incorrectly represent the graph of a function at or near points of discontinuity where the function is undefined.

You can determine limits at infinity as well.

```
g := x -> sqrt(x^2 - 4*x) - x;
limit(g(x), x = infinity);
```

Here you are looking at $\lim_{x \to \infty} g(x)$. Try rationalizing the numerator using

$$\sqrt{x^2 - 4x} + x$$

to check the result displayed by Maple.

Use the plot *command to examine the behavior of the function for large values of* x.

plot(g, 0..100);
The number 100 is not a very large value for x, but the graph gives you a feel for the behavior of the function.

You can use much larger values of x *to obtain more information about the function.*

plot(g, 100..1000);
The function seems to be flattening out. What value is the function approaching for values of x near 1000? Does this agree with the value given by Maple and the value you calculated?

You can examine piecewise defined functions at the points where their rules change. At these points, you use Maple's ability to determine left and right limits.

Start by defining the function and graphing it.

f := proc(x) if x < 0 then x - 1 else
    x^2 fi end;
plot(f(x), x = -2..2);
Notice that an error occurs. The **plot** command is unable to evaluate this function in this form. Maple can, however, graph this function.

You can use just f *in the* plot *command.*

plot(f, -2..2);
Now Maple graphs the function. The discontinuity is clearly shown at $x = 0$. Can you use open and closed dots to indicate accurately the behavior of the function at $x = 0$?

You can use one-sided limits to determine

$\lim\limits_{x \to 0^-} f(x)$ *and* $\lim\limits_{x \to 0^+} f(x)$

for

$$f(x) = \begin{cases} x - 1 & \text{for } x < 0 \\ x^2 & \text{for } x \geq 0 \end{cases}.$$

limit(x - 1, x = 0, left);
limit(x^2, x = 0, right);
Does this function have a limit at $x = 0$? An error message stating **cannot evaluate boolean** occurs if you try
limit(f(x), x, x = 0).

Derivatives of Functions

Maple can differentiate most elementary functions.

The differentiation command is `diff`.

`diff(3*x^4 - 4*x^2 - 5, x);`
You should have no problem checking the displayed result. Notice that it is necessary to indicate that you are differentiating the expression that defines a function with respect to x.

You can differentiate the quotient of functions.

`diff((x + 1)^2 / (x^2 + 2*x)^2, x);`
The function being differentiated is

$$\frac{(x+1)^2}{(x^2+2x)^2}$$

The result can be simplified.

Investigating a Function Using `plot` and `diff`

You can investigate the behavior of a function by graphing it. The information you garner from graphs of the function in conjunction with the zeros of its first and second derivatives can give you accurate approximations of the maxima, minima, and points of inflection. Consider the function:

$$f(x) = \frac{x+2}{(3 + (x^2 + 1)^3)}$$

You begin by entering the function as an expression.

`f := (x + 2)/(3 + (x^2 + 1)^3);`
The parentheses are needed in the numerator and denominator to ensure that Maple performs the operations in the order intended. This is a difficult function to investigate using pencil-and-paper methods.

Graph the function.

`plot(f, -5..5);`
Notice that the function seems to appear in the vicinity of -2 and seems to disappear in the vicinity of 2 and that the maximum value is less than 1.

Regraph the function with your own values for x and y.

```
plot(f, x=-5..5, y=-0.1..0.1);
```
The y values are indicated as -0.1 and 0.1 to avoid the possible confusion that might occur with multiple dots between the numbers. You can see now that the graph crosses the x-axis near -2. You should click on the Maple Sessions window rather than closing the Plot window. DOS users will have to handle this part differently. See below.

You can determine where maxima, minima, and points of inflection occur by looking at the derivatives of $f(x)$.

```
d := diff(f, x);
simplify(d);
fsolve(numer("));
```
These three statements give the first derivative and the approximate zeros for the numerator of $f'(x)$. These are, of course, the zeros of $f'(x)$. Referring back to the last graph by clicking on the last Plot window, you can see that the negative number is the position of the minimum for the function. DOS users—use the uparrow key to recall the last plot statement and redraw the graph.

You can more clearly see the behavior of the function to the left of zero by choosing appropriate x and y values.

```
plot(f, x=-4..0, y=-0.01..0.01);
```
The zeros before the decimal points are important. Is the minimum more discernible now? Does there appear to be a point of inflection to the left of this minimum?

Now for the points of inflection.

```
diff(d, x);
simplify(");
fsolve(numer("));
```
The expression d is the rule for $f'(x)$. Thus, the expression **diff(d, x)** is the second derivative of $f(x)$. Once again, you solve the numerator of the second derivative to reduce the amount of work that Maple has to do. This can cause problems if the denominator has real zeros.

Vertical tangent lines and cusps have not been considered here because the graphs did not seem to indicate them. How would you check to make sure there are no vertical tangent lines? Also, you might wish to know about horizontal and vertical asymptotes. You can investigate these using the **limit** command.

You have been working with a very difficult function whose most interesting behavior occurs very close to the x-axis. This would be very difficult to discover using pencil-and-paper methods. However, the combination of refining graphs and considering derivatives provides tools to allow you to understand thoroughly the behavior of the function. The analytical skills you can develop investigating such difficult functions with Maple will pay big dividends for you in any quantitative work you do in the future.

Integrals of Functions

Maple can determine indefinite and definite integrals of functions. Maple also has the ability to approximate definite integrals for functions whose antiderivative cannot be determined.

Integrals of polynomials are straightforward.

```
int(3*x^4 - 2*x, x);
```
You must indicate the variable of integration as with differentiation. Can you check the result displayed? *Hint:* Do you remember the **diff** command? Notice that a specific antiderivative is given, not the most general antiderivative. The constant of integration is not displayed.

Maple can integrate trigonometric functions such as $f(x) = \sec^4(x)$.

```
int(sec(x)^4, x);
```
Notice the placement of the exponent. Can you check the result?

Maple can perform integration by parts.

```
int(x^3 * ln(x), x);
```
You can use **diff(", x);** to check the result.

You can compute more complicated integrals such as
$$\int \frac{x^2}{\sqrt{x^2 - 9}} dx.$$

```
int(x^2 / sqrt(x^2 - 9), x);
```
Be careful when you enter this expression.

You can check this result.

```
diff(", x);
```
This does not look anything like the original integrand.

You can, however, simplify this expression.

```
simplify(");
```

Check carefully to make sure this matches the original integrand.

Finally, you can integrate functions such as
$$\frac{x^2 + 2x + 1}{(x^2 + 1)^2(x - 2)}.$$

```
int((x^2+2*x+1)/((x^2+1)^2*(x - 2)), x);
```

Clearly, you need to be careful when entering long expressions on one line. Remember, when you check this result, you may need to use the **simplify** command.

Definite Integrals of Functions of One-Variable

You can also evaluate definite integrals such as
$$\int_0^{\pi/2} \sin(x)dx.$$

```
int(sin(x), x = 0..Pi/2);
```

Once again, notice the double dots between **0** and **Pi/2**. The result of this computation is 1 so that the graph of one hump of the sine function is 2. MapleV is able to evaluate many definite integrals exactly, as in this case. However, you should be aware that not all definite integrals can be computed exactly by paper-and-pencil techniques or by using computer software.

The student package for calculus, available with Maple, allows you to determine the solutions to certain problems in a step-by-step fashion rather than by simply applying a single Maple command. This is very useful when you wish to learn the solution process or to see more deeply into the structure of a problem.

To use the student package for calculus, you must first load it.

```
with(student);
```

This makes the commands in this package available.

Let's look at an indefinite integral that can be evaluated using the integration by parts method.

```
Int(x*sin(x), x);
```

This is the integral $\int x \sin(x)\, dx$. The **Int** operator with a capital **I** creates the integral but does not evaluate it.

Apply the integration by parts command.

```
intparts(", x);
```

The previous output (**"**) is integrated by parts. The **x** in the Maple statement is your choice of which factor in the integrand is to be differentiated. You can see that the result contains an integral that is standard and easily integrated.

You can investigate more complicated integrals.

```
intparts(Int(x^2*exp(2*x),x),x^2);
```
This is the integral $\int x^2 e^{2x}\,dx$. The result includes an integral that appears simpler than the original. This should encourage you to apply the integration by parts method again.

Apply the method again and choose the expression that is to be differentiated.

```
intparts(Int(x*exp(2*x),x),x);
```
You should be able to recognize the integral in this result. The factor 2 needs to be inserted in the integrand (with appropriate adjustment) to make it a standard form. You can now finish the problem. You will notice that the signs and multipliers that result from the integration by parts method need to be accounted for. You may find it easier to do this yourself by hand.

The student package also allows you to go through the step-by-step procedure used in the trig substitution method.

You start with an integral.

```
Int(x^3*sqrt(1 - x^2), x);
```
Notice again that the integral is displayed and has not been evaluated. That is the difference between the **Int** and **int** commands. As you saw before, **int** evaluates integrals whereas **Int** gives the unevaluated form of the integral.

You decide to use trig substitution.

```
changevar(x = sin(v), ", v);
```
The substitution of $\sin(v)$ for x is made in $\int x^3\sqrt{1-x^2}\,dx$ and displayed.

Trig identities are necessary with this method.

```
powsubs(1 - sin(v)^2 = cos(v)^2, ");
```
The **powsubs** command allows you to replace $1-\sin^2(v)$ with $\cos^2(v)$ in the last expression (").

Since the sine function is raised to an odd power, you can factor a sin(v) *out and change the remaining even powers of sine to cosine.*

```
powsubs(sin(v)^2 = 1 - cos(v)^2, ");
```
You may already see the power forms in the displayed results.

Multiplying will make the power form more obvious.

```
expand(");
```

The value *command in the student package finds the value of the unevaluated expression.*

```
value(");
```
The evaluated integral is displayed but is not in terms of x. You can use the **subs** command to accomplish this but it is probably just as easy for you to perform the substitution by hand.

Investigating a Practical Problem

You can use Maple to assist you in solving real world problems that require extensive and complicated symbolic computations. For example, suppose you wish to determine the longest ladder that can fit around a right angle turn in a hallway if it is held horizontal. The diagram below summarizes the information you are given to solve this problem.

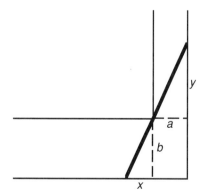

The widths of the two corridors of the hallway are labeled a and b. Using the two small right triangles, the length of the ladder is $\sqrt{(x+a)^2 + (y+b)^2}$ and, from similarity, the equation $\dfrac{y}{a} = \dfrac{b}{x}$ is satisfied by the labeled lengths.

Assign the length of the ladder to the variable `ladder`.

```
ladder := sqrt((x+a)^2+(y+b)^2);
```
In this expression, x and y are variables and a and b are constants or parameters that describe a particular hallway. The expression defines a function of two variables.

Enter the similarity based equation.

```
y/a = b/x;
```
This equation will be used to find y as a function of x.

Solve for y in terms of x.

```
solve(", y);
```
With this information, you can transform the expression for the length of the ladder from and expression in the two variables x and y into an expression in the single variable x.

Substitute the value of y in the `ladder` *expression.*

```
ladder := subs(y = ", ladder);
```
This result is an expression in one variable and defines a function of x with a and b as two constants or parameters.

Determine the extrema of this function by finding its critical points.

```
diff(ladder,x);
simplify(");
solve(numer(") = 0,x);
```
The single critical point (the other solutions are complex) is a surprisingly simple expression in the parameters a and b. But is this value for x a maximum or a minimum?

Investigate the second derivative.

```
diff(ladder, x$2);
simplify(");
```
The $x\$2$ calls for the second derivative. You can see that the second derivative is always positive by inspection. Thus, the ladder function is always concave up and the critical point gives rise to an absolute minimum for the function. How can this be?

You should look carefully at the diagram and how ladders of various lengths would appear in it. Each ladder touches at the corner of the turn and at the walls in both corridors.

The maximum length of a ladder that can negotiate the turn is the minimum length of ladders that can touch the corner and at the same time touch a wall in each corridor. You can determine particular ladder length values for given values of a and b by using the **subs** command with the expression for the critical point.

It should be noted that a different labeling of the diagram will lead to equations that are not so easily solved by Maple. For example, the labeling in the following diagram will lead to an expression in x or y, the zeros of whose derivative cannot be found exactly by Maple.

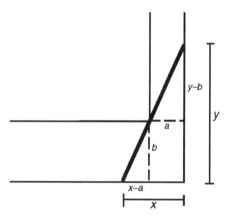

Also, since parameters a and b are used, numerical approximation techniques will not work either. Thus, you can use Maple to help you solve problems, but at each step you must think and consider ways to make a difficult computation less arduous both for yourself (with paper-and-pencil methods) and the computer software that you are using.

You may wish to redo the Maple computations with fixed values for the corridor widths. For example, you might set $a = 3$ ft and $b = 4$ ft to get a feel for how the computations work. You may also wish to check the reasonableness of your analysis with corridors of equal length where $a = b$.

3.2 Calculus II

Series

You can find the value of sums such as $\sum_{i=1}^{5} i^2$, create a Taylor series for a differentiable function, and create power series using recursive formulas.

You can sum a finite series.

```
sum(k^2, k = 1..5);
```
Here the function is the first argument and the range of the index values is second.

The last value for the index can be a variable.

```
sum(k^2, k = 1..n);
```
Notice that this formula is equivalent to the standard one you encountered in precalculus mathematics or in the introduction to integration in your calculus course.

You can find some infinite sums as well.

```
sum(1/2^k, k = 1..infinity);
```
Since this is a geometric series, Maple can determine the sum.

Taylor Series

A function that is infinitely differentiable can be represented by a infinite series called a Taylor series. The **taylor** command displays a specified number of terms of the Taylor series expanded about a specified value.

The function e^x is infinitely differentiable.

```
taylor(exp(x), x = 0, 5);
```
Note the comma after 0. The Taylor series is expanded about $x = 0$, and the first five terms are displayed. Notice the last term displayed. **O(5)** gives an indication of the degree to which the finite polynomial is inexact.

The Taylor series for more complicated functions can easily be obtained.

```
s := taylor(exp(-x)*cos(x), x = 2, 7);
```
Here the Taylor series is expanded about $x = 2$ and seven terms are requested. Notice that the last term displayed is **O(7)**, indicating the degree of accuracy of the approxi-

mation. You may wish to graph the function and its approximation as a visual check of how well the function is approximated.

Graph the function and its Taylor series approximation.

```
sp := convert(s, polynom);
plot({exp(-x)*cos(x),sp},x=-2..5);
```

Here, **convert** converts the Taylor series approximation into a polynomial by eliminating the order term. Notice that the graphs diverge to the left of 1 and to the right of 4.

Functions of Several Variables

Maple can determine the limit of a function of several variables. Maple can also find partial derivatives and multiple integrals of functions of several variables.

Derivatives of Functions of Several Variables

You can find the partial derivatives of $f(x,y) = x^2y^3 + e^x + \ln(y)$.

```
diff(x^2*y^3 + exp(x) + ln(y), x);
```

This yields the partial derivative of the function with respect to x:

$$\frac{\partial f(x,y)}{\partial x}$$

The partial derivative of $f(x,y) = x^2y^3 + e^x + \ln(y)$ *with respect to* y *can also be computed.*

```
diff(x^2*y^3 + exp(x) + ln(y), y);
```

This yields the partial derivative of the function with respect to y:

$$\frac{\partial f(x,y)}{\partial y}$$

As you might expect, you can find second partial derivatives.

```
diff(diff(x^2*y^3 + cos(x)*sin(y),x),y);
```

This is

$$\frac{\partial^2 f(x,y)}{\partial x \partial y}$$

A shorthand version of this command is:

```
diff(x^2*y^3 + cos(x)*sin(y),x,y).
```

You can determine other partial derivatives in the same way.

Applications of Partial Derivatives

Partial derivatives are used to compute the divergence and curl, two frequently used tools in science and engineering. A vector function **F** of three variables can be described by $\mathbf{F}(x, y, z) = M(x, y, z)\mathbf{i} + N(x, y, z)\mathbf{j} + P(x, y, z)\mathbf{k}$, where **i**, **j**, and **k** are the unit vectors in the positive x, y, and z directions, respectively. If M, N, and P are differentiable functions, then the divergence of **F** is the scalar function

$$\text{div}\mathbf{F} = \frac{\partial M}{\partial x} + \frac{\partial N}{\partial y} + \frac{\partial P}{\partial z},$$

and the curl of **F** is the vector function

$$\text{curl}\mathbf{F} = \left(\frac{\partial P}{\partial y} - \frac{\partial N}{\partial z}\right)\mathbf{i} + \left(\frac{\partial M}{\partial z} - \frac{\partial P}{\partial x}\right)\mathbf{j}$$
$$+ \left(\frac{\partial N}{\partial x} - \frac{\partial M}{\partial y}\right)\mathbf{k}$$

Maple allows you to obtain these results easily. Consider finding the divergence and curl of the vector function $\mathbf{F}(x, y, z) = xy\sin(z)\mathbf{i} + x^2\cos(y)\mathbf{j} + z\sqrt{xy}\mathbf{k}$. In order to involk the divergence and curl commands, you need to load the linear algebra package, **linalg**. Linear algebra is covered in much greater detail later in the book.

First load the linear algebra package.

```
with(linalg);
```
The commands available in this package are displayed.

Enter the components of the function as a list.

```
F := [x*y*sin(z),x^2*cos(y),z*sqrt(x*y)];
```

Next enter the vector with respect to which the divergence and curl will be taken.

```
v := [x, y, z];
```
This vector is also entered as a list.

Now for the divergence of **F**.	`diverge(F, v);` The divergence of **F** is displayed.
Finally, you can compute the curl of **F**.	`curl(F, v);` The curl of **F** is displayed. You can see how rapidly Maple returns these results. You can also see how partial derivatives are used in the definitions of two commonly used functions in the field of science and engineering.

3.3 3D Plotting and Functions of Several Variables

This section introduces three-dimensional graphing. You will learn how to represent graphs in a variety of ways.

Graphing Functions of Two Variables

Graphing the function $f(x, y) = x^2 + y^2$.	`plot3d(x^2 + y^2, x = -3..3, y = -3..3);` Here, the x and y intervals refer to the rectangle of domain values for the function defined by the expression $x^2 + y^2$. The range values (for z) are not specified and they will be computed automatically by Maple. The figure graphed is part of an elliptic paraboloid.

Use the `view` *option to specify the range.*	`plot3d(x^2 + y^2, x = -3..3, y = -3..3,`
	`view = 0..10);`
	The interval of range values for the vertical dimension is $[0, 10]$. The plot looks substantially different.
	    Let's explore the amazing features of the Maple3d plot window. Several menus at the top of the window can be used to change the way the plot is displayed. But first you will adjust the viewing angle using the figure rotation feature. DOS users—use the selections at the bottom of the screen and the Menu made available by the F10 key to accomplish the following tasks.

Let's rotate the figure forward.	*Place the arrow cursor on the figure. Hold down the mouse button (the left button on the IBM and the Sun) and notice that a box appears. Move the cursor straight up about an inch by dragging the mouse and release the mouse button.*

Press Enter to redraw the figure.

On the Sun, click on the **Plot** option box to redraw the figure. Notice that you are seeing mostly the outside of the figure with a glimpse of the interior of the figure.

Change the Style *and* Axes.

*Pull down the **Style** menu and select **Wireframe**. Then select **Boxed** from the **Axes** menu. Press Enter to redraw the figure.*

Three-dimensional plots are often difficult to display in a meaningful way. You will frequently need to adjust the way a figure is displayed to get a good visualization of the behavior of the function. You are encouraged to use the numerous options available in three-dimensional graphing as you investigate two-variable functions.

Display the axes.

*Select **Normal** from the **Axes** menu and redraw the figure.*

The three axes along with units are displayed.

Let's look at one more view.

*Select **Constrained** under the **Projection** menu and **Boxed** under the **Axes** menu. Redraw the figure.*

The figure changes substantially and is now well inside its box boundaries. As you can see, these options can change the figure dramatically.

Let's plot another function.

```
plot3d(x^2 - y^2, x = -3..3, y = -3..3);
```

This figure is called a saddle for obvious reasons. You should rotate it to see which views give you the most information. You may wish to select **Wireframe** under the **Style** menu to speed up the plots. Also, try **Patch** under **Style** if operating under DOS. The constrained view of this figure may give you some additional insight about the function as well.

One last plot.

```
plot3d(x*exp(-x^2 - y^2), x = -2..2,
y = -2..2);
```

This figure is very interesting to play with. Enjoy!

Multiple Integrals of Functions of Several Variables

You can integrate

$$\int_0^1 \int_0^{\sqrt{1-x}} xy^2 dy\, dx.$$

```
int(int(x*y^2,y=0..sqrt(1-x)),x = 0..1);
```

The fraction displayed is the exact value of this double integral. Triple integrals follow this same form.

An Application Using Multiple Integrals

You can find the center of mass of a plane lamina using multiple integrals. For example, suppose you wish to find the center of mass of a plane lamina in the shape of the region described by

$$0 \le x \le 4, \quad 0 \le y \le \sqrt{x},$$

if the density is given by $\rho(x,y) = xy$. You need to recall that the center of mass, $(\bar{x}, \bar{y})$, is

$$\bar{x} = \frac{M_y}{M} \quad and \quad \bar{y} = \frac{M_x}{M}$$

where M is the total mass of the lamina and M_x and M_y are the moments of the region with respect to the x-axis and y-axis, respectively.

You start by finding the total mass, M.

```
M:=int(int(x*y, y=0..sqrt(x)),x=0..4);
```

This is the double integral $M = \displaystyle\int_0^4 \int_0^{\sqrt{x}} xy\, dy\, dx$.

Next find the moment of the region about the x-axis, M_x.

```
Mx:=int(int(x*y^2,y=0..sqrt(x)),x=0..4);
```

This is the double integral $M_x = \displaystyle\int_0^4 \int_0^{\sqrt{x}} y(xy)\, dy\, dx$.

Now for the moment of the region about the y-axis, M_y.

```
My:=int(int(x^2*y,y=0..sqrt(x)),x=0..4);
```

This is the double integral $M_y = \displaystyle\int_0^4 \int_0^{\sqrt{x}} x(xy)\, dy\, dx$.

You can now calculate the center of mass, $(\bar{x}, \bar{y})$.

```
xbar := My/M;
ybar := Mx/M;
```

The center of mass is found to be $(\bar{x}, \bar{y}) = (3, \frac{8}{7})$.

Moments of inertia of a plane lamina and the radius of gyration about an axis can be found in a similar fashion. You need but recall the formulas and perform the appropriate integrations.

Limits of a Function of Several Variables

You can take the limit of a function of two (or more) variables as (x, y) approaches (a, b).

```
limit((x^2 - y^2)/(x^2 + y^2), {x=0,y=0});
```

This is

$$\lim_{(x,y)\to(0,0)} \frac{x^2 - y^2}{x^2 + y^2}$$

Maple returns "undefined". The limit does not exist since the limit along the y-axis is different from the limit along the x-axis.

Another example involving a limit at infinity.

```
limit((x + 1/y), {x = 0, y = infinity});
```

This is

$$\lim_{(x,y)\to(0,\infty)} x + \frac{1}{y}$$

The limit is 0.

Extensions

Maple Calculus Commands

You can differentiate, integrate, and take limits of functions of one or more variables using the **diff**, **int**, and **limit** commands in Maple. The **simplify** command is often useful in displaying results in a more readable form. The **plot** command can be used to obtain graphic information about a function. This graphic information in conjunction with the first and second derivatives and the **fsolve** command can be useful in investigating the behavior of a function.

You can load the student package for calculus using the **with** command. You can use the **intparts** command in this package to become familiar with the integration by parts technique.

Infinite sums can be investigated using the **sum** command. Taylor series can be obtained using the **taylor** command.

The student **Package**

The **student** package contains a variety of commands useful in a calculus course. The **student** package commands allow you to work through the details of a solution to a problem but do not solve the problem directly. The **Int** command sets up an integral that can then be manipulated with such commands as **intparts, changevar,** and **powsubs** to reformulate the original integral into a recognizable form. The **powsubs** command substitutes one expression for another in a given expression and is used in trig substitutions in integrals. The **value** command evaluates unevaluated expressions.

Additional Activities

Explore the following functions making use of the **diff, int, limit, plot, simplify, fsolve, numer, denom,** and **taylor** commands.

Accurately locate any maximum and minimum values and points of inflection, and draw a careful sketch of each of the following functions.

1. $f(x) = x^{2/3} - \frac{1}{5}x^{5/3}$

2. $f(x) = \frac{x^2 + x - 2}{x^2}$

3. $f(x) = \frac{\sqrt[3]{1 - x}}{1 + x^2}$

Evaluate the following limits.

4. $\displaystyle \lim_{x \to \infty} \frac{2x^2 - 1}{x^2 + 3}$

5. $\displaystyle \lim_{x \to -1} \frac{x^2 - 2x + 1}{2x^2 - x - 3}$

6. $\displaystyle \lim_{(x,y) \to (1,3)} x + y^2$

7. $\displaystyle \lim_{(x,y) \to (2,1)} \frac{xy - y}{2x + 1}$

Perform the following integrations.

8. $\displaystyle \int x \ln x \, dx$

9. $\displaystyle \int \frac{dx}{\sqrt{x^2 - 4}}$

10. $\displaystyle \int \frac{2x^2 + 19x - 45}{x^3 - 2x^2 - 5x + 6} \, dx$

11. $\displaystyle \int \frac{dx}{4 \sin x - 3 \cos x}$

12. $\displaystyle \int e^{-x} \cos x \, dx$

13. $\displaystyle \int_0^1 \int_0^{\sqrt{x}} y e^{x^2} \, dy \, dx$

Find and graph the first five terms of the Taylor series expansion about the indicated value a for each of the following.

14. $\cos x, \quad a = 0$

15. $\sin x, \quad a = 0$

16. $\cos x, \quad a = 1$

17. $x \sin 2x, \quad a = 0$

18. e^x, $a = 0$

For the given functions, show that $\dfrac{\partial^2 f(x,y)}{\partial x \partial y} = \dfrac{\partial^2 f(x,y)}{\partial y \partial x}$.

19. $f(x,y) = x^2 y^3 + \cos x \sin y$

20. $f(x,y) = 2^x y^2$

4 Linear Algebra

The `linalg` Package

The linear algebra package, called **`linalg`**, contains all the standard basic functions of linear algebra together with a number of special-purpose commands.

The first order of business is to load the **`linalg`** package. This has to be done once at the beginning of each session or if all variables are cleared (possible with some versions but not others).

You use the with *command to load the package.*

```
with(linalg);
```
The names of the commands and functions that are loaded are listed on the screen: Don't be surprised if you are unfamiliar with many of them; linear algebra has been around for a long time and it has many specialized functions.

The help system covers the new commands in the package.

```
?linalg
```
You can use Maple's help system at any time to remind yourself of the names of the **`linalg`** commands or to get help on their use. Alternately, you can call up the help system with the **`help`** command, as in **`help(linalg)`**.

4.1 Matrices and Vectors

In computer science, matrices are also called arrays. In Maple, a matrix is a two-dimensional array with entries indexed by two subscripts, as in $M = \begin{bmatrix} m_{11} & m_{12} \\ m_{21} & m_{22} \end{bmatrix}$. One-dimensional arrays are called *vectors* in Maple; the entries of a vector are indexed by a single subscript, as in $\mathbf{v} = (v_1, v_2)$.

You enter a matrix with the `matrix` *command. NOTE to Windows Users: To get a new line without sending anything to Maple, press SHIFT ENTER instead of ENTER.*

```
M:=matrix([
[1/2,2/3,3/4,4/5],
[5/6,6/7,7/7,8/9],
[9/10,10/11,11/12,12/13]
]);
```

In Maple, a bracketed sequence [...] of entries is a *list*. Here, the argument to the matrix command is a *list*, the entries of which are again lists. The inside lists are the rows of the matrix, in order. For example, the second row of the matrix M has entries 5/6, 6/7, 7/7, 8/9. You will encounter other forms of the **matrix** command later; this one is the *list-of-lists* form.

If you enter a matrix name, Maple simply echoes the name.

```
M;
```

Maple treats matrix names and function names analogously; when you enter the name of a matrix or function, the definition is not displayed.

You use the `evalm` *command to print the matrix in standard rectangular form.*

```
evalm(M);
```

Think of **evalm** as meaning "evaluate to a matrix."

You can also access the individual entries of the matrix.

```
M[2,3];
```

This returns the entry in the second row and third column. Notice that square brackets are used here whereas parentheses are used for functions.

The notation $M[i,j]$ is "computer subscript notation." Some texts use M_{ij} for the (i,j)-entry of M; others denote it by m_{ij}. Maple uses $M[i,j]$.

You can edit a matrix by reassigning its entries.

```
M[2,3]:=7/8;
```

Here the (2,3)-entry of the matrix M is changed to 7/8. You might wish to verify the change to M with the **evalm** command.

If a matrix has only one row or column, it may appear one-dimensional but in Maple it is still necessary to address the entries with two subscripts.

Consider the row and column matrices R and C.

```
R:=matrix([[1,2,3,4]]);
C:=matrix([[4],[3],[1],[1]]);
```

Both R and C appear one-dimensional but, as they are matrices, they are two-dimensional.

You have to use two subscripts to address the entries of any matrix—even one which appears one-dimensional.

```
C[1];
R[1];
```

Some texts use only one subscript to address the entries of row and column matrices; the error messages generated here demonstrate that Maple does not.

Use the vector *command to generate an array in which the entries are addressed by a single subscript.*

```
v:=vector([4,3,1,1]);
v[3];
```

In this case, the argument to vector is a single list. Note that Maple does not consider **v** to be a matrix; the data structure **v** is perhaps best thought of as the ordered 4-tuple $v = (v[1], v[2], v[3], v[4])$.

Maple is consistent in its response to array names.

```
v;
```

Consistency is a virtue.

You edit vectors, like matrices, by reassigning the components.

```
v[3]:=2;
```

Modifying the components of a one-dimensional array is analogous to modifying the entries of a two-dimensional array, except that only one index is used.

The evalm *command works for vectors as it does for matrices.*

```
evalm(v);
```

Note that Maple prints the vector with commas between its entries; this is not so of matrices of modest width.

Solving Systems of Linear Equations

Solving systems of linear equations is a straightforward but time-consuming, error-prone task for which Maple is well equipped. You have already used the solve command in Chapter 2. You can use the solve command to solve any system of equations. Here is a typical system of linear equations.

$$\begin{cases} 3x_2 - 4x_3 + \dfrac{5}{3}x_4 = \dfrac{23}{12} \\[2mm] 2x_1 + 7x_2 + \dfrac{4}{3}x_3 + 3x_4 = \dfrac{41}{4} \\[2mm] \dfrac{1}{2}x_1 - 3x_2 + 2x_3 + \dfrac{13}{3}x_4 = \dfrac{41}{12} \\[2mm] \dfrac{7}{6}x_1 + \dfrac{7}{3}x_2 - \dfrac{14}{9}x_3 + 7x_4 = \dfrac{35}{4} \end{cases}$$

This enters the system into the session.

```
Sys:=
{3*x2-4*x3+5/3*x4=23/12,
2*x1+7*x2+4/3*x3+3*x4=41/4,
1/2*x1-3*x2+2*x3+13/3*x4=41/12,
7/6*x1+7/3*x2-14/9*x3+7*x4=35/4};
```

Note that commas are used to separate the equations and that the sequence of equations is enclosed in set braces.

You can now use solve *to find the solutions.*

```
solve(Sys,{x1,x2,x3,x4});
```

Maple reports all solutions that it finds—in this case all that exist.

Notice that $x1$, $x2$, $x3$, $x4$ are used here rather than $x[1]$, $x[2]$, $x[3]$, $x[4]$. This style of "subscripting" seems more natural in some circumstances, but either style can be used.

You may find that you prefer to solve a linear system by using Gaussian elimination on the augmented matrix of the system.

```
AM:=
matrix([
[0,3,-4,5/3,23/12],
[2,7,4/3,3,41/4],
[1/2,-3,2,13/3,41/12],
[7/6,7/3,-14/9,7,35/4]
]);
```

Entering the augmented matrix of the system requires significantly fewer keystrokes than entering the equations.

If you like, you can step through the Gaussian elimination procedure using the Maple commands **swaprow**, **mulrow**, and **addrow**.

You use swaprow *to interchange the order of two rows.*

SR:=swaprow(AM,1,2);
Note that the matrix name is the first argument to the command, followed by the numbers of the rows to be swapped.

You use mulrow *to multiply a scalar times a row.*

MR:=mulrow(SR,1,1/2);
The matrix name comes first, then the number of the row to be multiplied, and then the multiplier.

You use addrow *to add a multiple of one row to another row.*

AR:=addrow(MR,1,3,-1/2);
Again, the first argument is the matrix name and the second argument is the number of the row to be multiplied; this is then followed by the number of the row to which the multiple is to be added, and then by the multiplier.

Alternately, you can partially automate the reduction by using pivot *in place of* addrow.

pivot(AR,2,2);
The **pivot** command uses **addrow** to zero the entries above and below the specified entry. Here, the (2,2)-entry is specified. You can also limit the pivot operation to forward or backward pivoting. Use **?pivot** for more details.

Maple also provides several automated procedures for Gaussian elimination. There are a few restrictions placed on these routines; they work for matrices with entries that are quotients of polynomial expressions with rational coefficients but they will not work on matrices which contain, for example, square roots which are not rational. For such matrices with more general entries, you will have to fall back on manual row reduction with **addrow**, **mulrow**, **swaprow**, and **pivot**.

The gausselim *routine uses elementary row operations to do forward elimination.*

GE:=gausselim(AM);
The first nonzero entry of each row of the result is frequently called the pivotal entry. Note that the entries above the pivotal entries have not been zeroed.

Alternately, you can do fraction-free forward Gaussian elimination.

```
FF:=ffgausselim(AM);
```

As its name suggests, the **ffgausselim** command performs Gaussian elimination without introducing any fractions.

You can take the matrix AM all the way to reduced row echelon form with the gaussjord *command.*

```
RR:=gaussjord(AM);
```

The **gaussjord** command performs standard Gauss-Jordan elimination on the matrix to produce the reduced row echelon form.

Alternately, you can use the command rref *as a synonym for* gaussjord.

```
rref(AM);
```

You may prefer **rref** to **gaussjord** for its brevity. Of course, **rref** is an abbreviation of "reduced row echelon form."

The reduced row echelon form is preferred by many because it is unique and it makes the back substitution trivial. However, numerical analysts frequently prefer to do only forward elimination (as is done with the **gausselim** routine) and then perform back substitution.

While it is a fairly simple matter to complete the solution of the original system by back substitution using any of the matrices *GE*, *FF*, or *RR*, you may choose to have Maple do it for you. This is especially handy for larger systems, where back substitution can become quite time consuming.

Use backsub *to automate back substitution.*

```
backsub(FF);
```

The **backsub** command will work on any matrix in echelon form. Each of the routines **gausselim**, **ffgausselim**, and **rref** (**gaussjord**) leaves the matrix in a form suitable for **backsub**.

Note that the solution produced by **backsub** is completely general; all solutions are produced by varying the parameter which appears in the solution.

Sometimes you may need to solve several systems with the same coefficient matrix. This is a situation that frequently occurs in applications. For example, assume you need to solve the system **Sys**, with which you have been working, for each of a variety of "right-hand sides" including $c[1] = (23/12, 41/4, 41/12, 35/4)$, $c[2] = (5, 0, -13/4, 7/4)$, etc.

The augmented matrix, AM, of the first system has already been entered and the first system solved. An analogous procedure would suffice for the remaining systems. But if you think of AM as the matrix K of coefficients of the unknowns augmented by the column c of right-hand sides, then it would clearly be more efficient to extract K and reuse it with $c[2]$, etc. Maple has block-editing commands that allow you to do this.

You can extract the coefficient matrix from AM with the submatrix *command.*

```
K:=submatrix(AM,1..4,1..4);
```
The first argument to the **submatrix** command is the matrix name, followed by the row range and then the column range.

Augment K by each of the new right-hand sides, in turn, and then proceed as before.

```
c[2]:=vector([5,0,-13/4,7/4]);
AM:=augment(K,c[2]);
gausselim(AM);
backsub(");
```
Notice that here, $c[2]$ is used as a name for a vector. The ith component of $c[2]$ is $c[2][i]$. The use of this notation implicitly defines c as a *table*. Tables are related to arrays but are more general. The command **print(c)** will display the table. You can use **?table** for more information on tables.

You can determine all vectors g for which the system is consistent.

```
g:=vector([g1,g2,g3,g4]);
AM:=augment(K,g);
GE:=gausselim(AM,4);
```
The second argument to **gausselim** (4 in this case) stops the **gausselim** procedure from using pivots to the right of the specified column; this ensures that entries which are conditionally zero are not used as pivots.

In this case, the system is consistent if, and only if, the (4, 5)-entry of GE is 0.

```
g[1]:=solve(GE[4,5]=0,g1);
```
Clearly, setting $GE[4,5]=0$ is equivalent to letting $g[1]$ be the solution of the equation $GE[4,5]=0$.

The general solution vector is now **g**.	`evalm(g);` Any vector **v** for which the linear system with augmented matrix $[K,\mathbf{v}]$ is consistent is now of the same form as **g**: **v**[2], **v**[3] and **v**[4] are arbitrary and **v**[1] = **v**[4] − 1/3**v**[2] − **v**[3].
Proceeding as before, you can find the solution of the general (solvable) system.	`AM:=augment(K,g);` `GE:=gausselim(AM,4);`
Complete the process with `backsub`.	`backsub(");` The vector returned gives a formula for the coefficients of the solution.

Further Editing Commands

It is quite convenient at times to be able to use just a part of a matrix, or to piece vectors and matrices together to form new matrices, as when several linear systems with the same coefficient matrix are to be solved. In addition to the **submatrix** and **augment** commands, Maple provides several other useful "editing tools."

With the **submatrix** command, you specify the range of rows and columns you want to keep. To keep one row or column, it may be easier to use the **row** or **col** commands.

Use the `row` *command to pick out a single row.*	`r:=row(AM,3);` The **row** command returns a vector, in this case, the third row of AM.
Use the `col` *command to pick out a single column.*	`c:=col(AM,5);` The **col** command also returns a vector, in this case, the fifth column of AM.

Sometimes you may prefer to specify a range of rows or columns you want to delete. You can do this with the **delrows** and **delcols** commands.

Use `delrows` *or* `delcols` *to delete rows or columns.*	`delcols(AM,5..5);` The first argument to **delcols** is the name of the matrix;

the second argument is the range of columns to be deleted. Notice that this is equivalent to the less compact command **submatrix(AM,1..4,1..4)**. The **delrows** command works analogously to delete rows.

The stack *command allows vertical augmentation.*

```
stack(K,c,v);
```
Both **augment** and **stack** will accept any number of arguments. The arguments can be matrices or vectors or any combination thereof. The lengths of the vectors must all be the same and (for **stack**) equal to the width of any matrix arguments.

An Important Point on Names

On some occasions, it may be convenient to make a copy of an existing array and change a few of the entries. Here, you can use the **copy** command.

Use copy *to make a copy of an array.*

```
cc:=copy(c);
cc[1]:=0;
cc[1];
c[1];
```
Note that $c[1]$ is unchanged. In this case, cc is not the same matrix as c.

The result is quite different if the copy *command is not used.*

```
cc:=c;
cc[1]:=0;
c[1];
```
In this case, cc and c are two names for the same matrix.

Summary and Extensions

You use the **matrix** command to enter a matrix as in

$$A := \text{matrix}([[1,2],[3,4]])$$

You use the **vector** command to enter a vector as in

v:=vector([1, 2, 3, 4])

It is best to think of vectors as n-tuples in Maple. In particular, vectors are not matrices.

The (i,j)-entry of a matrix A is denoted $A[i,j]$.

The ith entry of a vector **v** is denoted **v**[i].

You use the assignment operator (:=) to change an entry of a matrix or vector. The command **A[2,3]:=4** assigns the value 4 to the (2, 3)-entry of a matrix A. The command **v[2]:=3** assigns the value 3 to the second component of **v**.

You use **evalm(E)** to have Maple evaluate and display the entries of a matrix or vector expression E.

Systems of linear equations can be solved by row reducing the augmented matrix of the system and applying **backsub**. The row reduction can be done step-by-step with **addrow**, **mulrow** and **swaprow**, or automated with **gausselim**, **ffgausselim** or **gaussjord (rref)**. Alternately, you can partially automate the reduction with **pivot**.

Maple provides several editing commands for matrices. The **augment**, **stack**, **submatrix**, **delrows** and **delcols** commands allow easy cutting and pasting of parts of matrices.

Related commands (use the help facility (?) for further information): **copyinto**, **minor**, **subvector**.

Additional Activities

1. Enter the 4×6 matrix $A = (a_{ij})$ where each entry is defined by $a_{ij} = i/(i+j)$.

2. Use **gausselim** and **backsub** to solve the linear system having the matrix A of Activity 1 as its augmented matrix.

3. Use the **delcols** and **augment** commands to replace the sixth column of the matrix A of Activity 1 by the vector $\mathbf{v} = (v_i) \in R^4$ defined by $v_i = i^{32}$; save the result as B.

4. Use **gausselim** and **backsub** to solve the linear system having the matrix B of Activity 3 as its augmented matrix.

5. Solve the following linear system by using row reduction and back substitution on its augmented matrix.

$$\begin{cases} 2x_1 + 3x_2 + 4x_3 + 5x_4 + 6x_5 + 7x_6 = 8 \\ 3x_1 + 3x_2 + 4x_3 + 5x_4 + 6x_5 + 7x_6 = 11 \\ 4x_1 + 4x_2 + 4x_3 + 5x_4 + 6x_5 + 7x_6 = 37 \\ 5x_1 + 5x_2 + 5x_3 + 5x_4 + 6x_5 + 7x_6 = 32 \end{cases}$$

6. Use **ffgausselim** and **backsub** to show that, if a $\neq$ b, then, for every x, the linear system with the following augmented matrix is consistent.

$$M = \begin{bmatrix} a & b & x \\ a+1 & b+1 & x+1 \\ a+2 & b+2 & x+2 \\ a+3 & b+3 & x+3 \\ a+4 & b+4 & x+4 \\ a+5 & b+5 & x+5 \\ a+6 & b+6 & x+6 \\ a+7 & b+7 & x+7 \\ a+8 & b+8 & x+8 \\ a+9 & b+9 & x+9 \end{bmatrix}$$

4.2 More on Matrices and Vectors

Matrix and Vector Arithmetic

The same arithmetic operation symbols are used on matrices as are used on numbers, except that the compound symbol "ampersand-star" (&*) is used for matrix multiplication and the **innerprod** command is used for matrix-vector and vector-matrix multiplication. The "star" symbol (*) is reserved for scalar multiplication.

The **evalm** command is applied to vector and matrix expressions to instruct Maple to evaluate them.

First things first: `with(linalg);`

One approach to matrix-vector arithmetic is to build the expression to be evaluated and then apply the **evalm** command.

You build an arithmetic matrix expression as you would any arithmetic expression, using "&*" for matrix products.

Consider the two matri-
ces A and M given here,
for example.

```
A:=matrix([
[0,3,-4,5],
[2,7,4/3,9],
[1/2,-3,2,13],
[4/3,7,-2/3,11/3]]);
M:=matrix([
[1,2],
[3,4],
[2,3],
[4,5]]);
```

The matrices can be entered all on one line or split at any convenient point.

Define the expression S from
A and M.

```
S:=A &* M;
```

Note that S is not evaluated. The spaces around **&*** are for emphasis in this first use of the symbol; spaces are not required in general, except that a space is required between **&*** and ".

You can use expressions
in defining other expres-
sions.

```
F:=A^2&*(2*M-S-M);
```

Note that the definition of F uses S.

Use evalm *to get the eval-*
uated form of the result.

```
evalm(F);
```

The result is given in standard rectangular form.

It is also possible to "wrap"
expressions with evalm,
using standard function
notation.

```
evalm(A&*M+S);
```

Sometimes you may want to manipulate an expression before calling **evalm**. At other times, you may just want the detailed answer straight-away. You will develop your own preferred style with **evalm**. Think of it as a "button" you push when you want to see the answer in full matrix form.

You can use print *to review the definition of an expression.*

print(F);

The **print** command does not cause a matrix expression involving arithmetic operators to be evaluated. (Some simplification may be performed.)

Following standard convention, Maple will sometimes use 0 *to denote the zero matrix.*

A-A;

When Maple returns the symbol **0** for the zero matrix, there is no way to determine the dimensions of the matrix except to look back at the dimensions of the matrices that produced the result. However, typically the answer gives all the information that is required.

You can treat a scalar as a scalar matrix for addition.

evalm(A+3);

Note that the scalar has been treated as if it were the matrix $S = (s_{ij})$ of the same size as M with

$$s_{ij} = \begin{cases} 3 & \text{if } i = j \\ 0 & \text{if } i \neq j \end{cases}.$$

You cannot, however, treat a scalar as a scalar matrix for multiplication.

evalm(3&*A);

Maple insists that * be used for multiplication by scalars.

One seeming exception to this rule is that 0 *can be used with* &*.

evalm(0&*A);

This is consistent with Maple's use of **0** as the zero matrix.

In most cases, you cannot use the scalar multiplication operator for matrix multiplication. A useful exception is for powers.

B:=evalm(A*A);

Maple evaluates products that use * without regard to the order of the terms. Since matrix multiplication is highly noncommutative, the designers of Maple have taken a conservative approach to allowing * for matrix products. However, order is not a problem for powers.

Maple can be very particular about the use of * *and* &*.

evalm(M&*3*S);

In this case, Maple uses left-to-right evaluation because &* and * have the same precedence. Since **M&*3** is an illegal expression, an error message results. Either use **evalm(M &*(3*S))** or relocate the scalar.

Vector addition and scalar multiplication are similar to the corresponding matrix operations.

```
v:=vector([-143/3,-47,-99/2,-31]);
evalm(2*v-k*v);
```
Here, k acts as a symbolic scalar because it has not been assigned a value.

You can use an array without naming it.

```
evalm(v+vector([3,7,-2/3,6]));
```
Matrices can be used in the same "on the fly" manner.

You can add a scalar to a vector but the result is not analogous to adding it to a matrix.

```
w:=evalm(v+3);
```
Here, the scalar has been treated as if it were the vector $\mathbf{s} = (3,3,3,3)$. Compare this result to the calculation of $M + 3$ above.

You can multiply matrices and vectors, in either order, using the innerprod *command. The result is a vector.*

```
innerprod(A,w);
innerprod(v,A);
```
In the first case, $\mathbf{w}$ is treated as if it were a column. In the second case, $\mathbf{v}$ is treated as if it were a row. This is a common extension to the definition of matrix multiplication.

You can use innerprod *with a vector on either end (or both or neither) and one or more matrices in the middle.*

```
innerprod(v,A,B,w);
```
Here, $\mathbf{v}$ is treated as if it were a row and $\mathbf{w}$ is treated as if it were a column. Of course, the computation will fail if the arrays are not of appropriate sizes to be multiplied—the length of $\mathbf{v}$ must agree with the height of A, the width of A with the height of B and the width of B with the length of $\mathbf{w}$. In this case, the result is a scalar.

The ampersand-star operation can also be used to compute the product $A\mathbf{w}$ of a matrix times a vector if the vector is on the right-hand side.

```
evalm(A&*w);
```
Vectors are treated as columns by the "ampersand-star" operation. The result is the same as **innerprod(A,w)**.

However, you cannot compute the product $\mathbf{v}A$ using the ampersand-star operation.

```
evalm(v&*A);
```
Vectors can only be used on the right side of the &* operation.

You can use the rather unappealing compound symbol "&*( )" for the identity matrix. But you might wish to `alias` *it to something else.*

```
alias(I=&*());
evalm(I&*A);
```

You can choose pretty much any name you wish for the identity in this way.

Notice that whenever you use the **alias** command, it returns a complete list of all the aliases in effect.

If you add the identity to another matrix, the effect is the same as adding the scalar `1`.

```
evalm(A+I);
```

The symbol "I" is initially aliased to the complex number i. Changing it simply causes Maple to respond $(-1)^{1/2}$ when it would have responded with "I", but you might prefer to use an alternate symbol, perhaps Id, for the identity matrix.

Other Forms of the Matrix and Vector Commands

Maple provides several alternative forms of the **matrix** and **vector** commands which you may find preferable in some cases.

The **matrix** command can be used in the form

matrix(m,n,spec)

where m and n are the number of rows and columns and *spec* is either a list of lists, or a list of entries, or a vector or a function of two variables.

The **vector** command can be used in the form

vector(m,spec)

where m is the number of entries and *spec* is either a list or a function of one variable.

You can specify m and n with the list-of-lists form of the `matrix` *command, if you wish.*

```
M:=matrix(2,3,[[1,2,3],[2,3,4]]);
```

As you now know, the specification of m and n is optional in this case.

You have similar latitude with the `vector` *command.*

```
v:=vector(3,[1,2,3]);
```

The specification of m is optional.

If you supply m and n, the third argument to the `matrix` *command can be a simple list, rather than a list of lists.*

`L:=matrix(4,2,[1,2,3,4,5,6,7,8]);`

This is the *list form* of the **matrix** command. When the matrix is formed, the list is broken into four groups of two entries. Many people prefer this form over the list of lists form for ease of data entry; others feel it lacks some of the naturalness of the list-of-lists form.

While a vector is not just a list, Maple will let you treat it as one in this one case.

`V:=matrix(3,1,v);`

This facility provides an easy way to convert a vector to a matrix. The order of the indices determines whether the result will be a row or a column. Type **?matrix** for a complete description of the use of vectors with **matrix**.

The third argument to the matrix command can also be a function of two variables.

`M:=matrix(7,9,(i,j) -> i/j);`

This is the function form of the matrix command. The third argument can be any function of two variables. In this case, the (i,j)-entry will be i/j. This is usually the quickest form to use when it is applicable.

Named functions, either built-in or user-defined, provide particularly easy matrix entry.

`Z:=matrix(3,3,0);`

For any rational number r, the function $x \to r$ is called r in Maple. Many texts use this convention for all numbers but Maple uses it only for rationals.

You can use the 0 function and "generic identity matrix" I to define identity matrices of a particular size.

`Ident[9]:=evalm(matrix(9,9,0)+I);`

Since it is so easy to create an identity matrix of any desired specific size, Maple does not have an identity matrix function. Note that the notation $I[9]$ cannot be used here—it would conflict with the aliasing of &*() to I.

You can also use a function form of the `vector` *command.*

`v:=vector(20,i->i^2);`

Here, the function is a function of only one variable. The result contains the squares of the first twenty natural numbers.

Diagonal matrices are easy to enter with the `diag` *command.*

`diag(1,5,2,4,3);`

This saves a lot of time over any other way of entering diagonal matrices.

The diag *command will also accept square matrices as arguments.*

```
diag(Z,1,2,matrix(3,3,1));
```
The result is what is called a *block diagonal matrix*. Note that **diag** is a synonym for **BlockDiagonal**. Both commands accept either "scalars" or square matrices.

If you specify m and n with the matrix *command, the third argument can also be omitted completely. The result is a symbolic matrix.*

```
S:=matrix(3,3);
evalm(S);
```
The (i,j)-entry of S is simply $S[i,j]$. Symbolic matrices are useful for general verification of matrix properties.

This creates a symbolic vector.

```
s:=vector(5);
evalm(s);
```
The ith component of **s** is **s**[i].

If you want to test an idea on numerical matrices, Maple will generate examples for you.

You can generate random integer matrices.

```
R:=randmatrix(20,20);
```
This generates a "pseudo-random" 20×20 matrix of integers.

You can test a hypothesis on a large matrix generated with the sparse *option.*

```
RS:=randmatrix(30,15,sparse);
```
The **sparse** option causes the matrix to have a lot of zero entries.

A Note on Arrays, Loops and Sequences

The loop is one of the most useful utilities in computing. In particular, loops are a great convenience for automating repetitive tasks.

The principal form of the loop in Maple is

for *count* **from** *start* **by** *inc* **to** *finish*
  do
    stufftodo
  od;

where *count* is a variable, *start* and *finish* are the beginning and ending values of *count*, and *inc* is the amount by which *count* is to be incremented on each repetition. The thing(s) to be done are listed between the **do** and **od** (and separated by semicolons if there is more than one). The "from clause" and "by clause" are optional; if omitted, the default values of *start* or *inc* are 1. Hence, "**for i to 3 do** ... **od**" is equivalent to "**for i from 1 by 1 to 3 do** ... **od**." The variable *count* can be an assigned or unassigned variable; in either case, it has the value *finish* +1 after the execution of the loop.

You can use loops to edit vectors or matrices.

```
v:=vector(27,i->1);
for i from 1 by 2 to 27
   do
      v[i]:=0
   od;
```

All entries of **v** with odd subscripts are changed to 0.

This verifies the change.

```
evalm(v);
```

Loops are also convenient for defining a collection of related structures.

```
for i from 1 to 9
   do
      v[i]:=vector(5,j -> j/i)
   od;
```

The counter retains a value after the execution of the loop.

```
i;
```

The fact that **i** has a value will cause no problems if you reuse it in a loop (the loop will reset it), but it might cause unpleasant surprises in other uses of **i**.

Note that Maple distinguishes between i *and* "'i'".

```
'i';
```

The value of `'i'` is the original, unassigned value of **i**— not the value assigned to **i**.

If you have no need for i *to retain its present assignment, it is best to unassign it.*

```
i:='i';
```

This has the effect of unassigning **i**. Note that you use regular (forward) quotes and not back quotes here.

Summary

The **matrix** command has a variety of forms.

The list form of the **matrix** command is generally quicker to use than the list-of-lists form. However, both forms have their proponents.

The function forms of the **matrix** and **vector** commands are probably the most efficient to use when they are applicable.

Symbolic matrices and vectors provide a means of verifying properties in general for low dimensions.

The **randmatrix** command generates pseudo-random matrices which are helpful for checking conjectures. A calculation with randomly generated numerical matrices is generally much quicker than the same calculation with symbolic matrices. Hence, **randmatrix** provides an attractive way to test hypotheses.

You can use the **diag** command to more easily enter diagonal (or block diagonal) matrices.

You can use the **innerprod** command to compute the product of a matrix and a vector, a vector and a matrix, a vector and a vector or a matrix and a matrix.

Related commands: array, **BlockDiagonal**, **companion**.

Additional Activities

1. Enter the 5×8 matrix A for which the (i,j)-entry is the quotient i/(i+j).

2. Find a function form for the definition of the matrix

$$A = \begin{bmatrix} 1 & 2 & 3 & 4 \\ 2 & 3 & 4 & 5 \\ 3 & 4 & 5 & 6 \\ 4 & 5 & 6 & 7 \\ 5 & 6 & 7 & 8 \end{bmatrix}.$$

3. Create the matrix

$$B = \begin{bmatrix} 1 & 2 \\ 2 & 3 \\ 4 & 5 \end{bmatrix}$$

by using the **submatrix** command on the matrix A of Activity 2.

4. Repeat Activity 3 using the **delrows** and **delcols** commands.

5. Use the function form of the **vector** command to enter the vector $\mathbf{v}$ in R^{20} with ith entry $\mathbf{i}^3$.

6. Use the function form of the **matrix** command to enter the 10×10 matrix $I_{10} = (\delta_{ij})$ where

$$\delta_{ij} = \begin{cases} 1 & \text{if } i = j \\ 0 & \text{if } i \neq j \end{cases}.$$

Let M be a random 10×10 matrix, and let A be the augmented matrix $[M, I_{10}]$. Row reduce A to reduced row echelon form $F = [L, R]$ and verify that the right-hand side R of F satisfies $RM = L$. Note that L is the reduced row echelon form of M.

7. Use **augment** and **rref** to show that every vector $\mathbf{x}$ in R^{10} is a multiple of the 10×10 matrix $M = (m_{ij})$ defined by $m_{ij} = \min(i, j)$.

8. Verify that the matrix

$$A = \begin{bmatrix} 1 & 2 & 3 & 4 \\ 2 & 3 & 4 & 5 \\ 3 & 4 & 5 & 6 \\ 4 & 5 & 6 & 7 \end{bmatrix}$$

satisfies the equation $A\mathbf{v} = (x, x+1, x+2, x+3)$, for every vector $\mathbf{v} = (3 + 2t - x + 2, -2s - 3t + x - 1, s, t)$.

9. Let

$$B = \begin{bmatrix} 1 & 2 & 3 & 4 & 5 \\ 2 & 3 & 4 & 5 & 1 \\ 3 & 4 & 5 & 1 & 2 \\ 4 & 5 & 1 & 2 & 3 \\ 5 & 1 & 2 & 3 & 4 \end{bmatrix}.$$

Evaluate the "matrix polynomial" $B^5 - 15B^4 - 25B^3 + 375B^2 + 125B - 1875$.

4.3 Basic Matrix and Vector Functions

The standard basic matrix and vector functions and commands are contained in the **linalg** package. In this section, you will see how they are used in the Maple environment.

At least you only have to do this once per session.

```
with(linalg);
alias(I=&*());
```

If, during a session, you find that Maple seems to not understand **linalg** commands, try a simple one like **matrix(1,1)**. If you have loaded the **linalg** package, Maple should respond **[?[1,1]]**. If Maple responds instead with **matrix(1,1)**, you probably have forgotten to load the **linalg** package. If loading **linalg** cures this problem, you may have to redefine any matrices or vectors you entered before loading the package.

You can compute the determinant of a matrix.

```
Max:=matrix(5,5,(i,j)->max(i,j));
det(Max);
```

You can compute the adjoint of a matrix—not a favorite pastime for many when done by hand.

```
Adj:=adjoint(Max);
```

The calculation of the adjoint of a 5×5 matrix requires the calculation of the determinants of 25 4×4 matrices.

If you wish to investigate the properties of a function like det, *you can work with symbolic matrices and vectors.*

```
S:=matrix(2,2);
det(S);
```

This gives the familiar formula for the determinant of a 2×2 matrix. You will not want to do this for matrices much larger than 4×4, however, due to the length of the output. (The formula for the determinant of a 5×5 has 120 terms, each a product of 5 terms; some machines may not have the resources to handle that. Try it.)

You can compute the inverse of an invertible matrix.

```
inverse(Max);
```
This can also be done with either **evalm(Max^(-1))** or **evalm(1/Max)**.

The simplicity of the inverse of the matrix Max may be bit of a surprise. You might like to experimentally determine a formula for the inverse of the $n \times n$ analogue of Max.

If you apply the inverse *command to a singular matrix, you get an error message.*

```
M:=matrix(5,5,(i,j)->i+j mod 2);
inverse(M);
```
Some programs return a "generalized inverse" if the regular inverse does not exist. Maple does not.

You can compute the inverse of a matrix with undefined entries.

```
inverse(S);
```
This gives the familiar formula for the inverse of an invertible 2×2 matrix. Like **det**, the **inverse** command requires substantial resources to compute inverses of larger symbolic matrices. Even if your machine has the resources to handle it, chances are you will not want to wait for a formula for the inverse of a 20×20 matrix.

You can check that the product of a matrix and its adjoint is the scalar matrix $\det(A)I$.

```
S:=matrix(4,4);
A:=adjoint(S);
alias(DetS=det(S));
evalm(A&*S);
```
Aliasing **DetS** to **det(S)** gives the output in simple form.

You can use inverse *to solve a linear system* $Mx = y$ *if the coefficient matrix is nonsingular.*

```
y:=vector([6,9,6,9,6]);
evalm(inverse(M)&*y);
```
Of course, this works only if M is invertible, which it is not in this case.

Alternately, you can use the linsolve *command to solve a linear system* $Mx = y$.

```
linsolve(M,y);
```
The **linsolve** command is somewhat analogous to multiplication on the left by the inverse, but does not require that the coefficient matrix be invertible.

This solves the linear system $Mx = y$ for x. In this case, there are many solutions. Notice that if $A = [M, \mathbf{v}]$ is the augmented matrix of the system, then **linsolve(M,v)** is analogous to **G:=backsub(gausselim(A))**. However, the solutions from the two procedures will not always have precisely the same form because the procedures use slightly different algorithms. Note that the components of the solution vector can be used to write y as a linear combination of the column vectors of M.

You can also use linsolve
to solve a matrix equation
$AX = B$ *for* X.

```
B:=augment(y,2*y);
X:=linsolve(M,B);
```
The second argument need not be a column matrix. Notice that the solution has been assigned the name X.

You may find the response
of linsolve *puzzling on*
occasions.

```
linsolve(M,Max);
```
All solutions found are reported—in this case, none.

You can compute the trans-
pose of a matrix.

```
transpose(X);
```
As can see, the rows and columns are interchanged.

The standard vector functions are also available.

You can compute the cross
product of vectors in R^3.

```
u:=vector([1,1,1]);
v:=vector([0,1,-1]);
w:=crossprod(u,v);
```
Recall that the cross product of two vectors is orthogonal to both.

You can compute the dot-
product of vectors.

```
dotprod(u,w);
```
Note that this is the same as **innerprod(v,w)** in this case, the result verifies that **w** is orthogonal to **v**.

You can compute the norm
of a vector.

```
norm(u,2);
```
Mathematicians use a variety of norms on vectors and matrices; this is the standard one $\|\mathbf{v}\| = \sqrt{\mathbf{v} \cdot \mathbf{v}}$. You might wish to compare the result of this command to the result of the command **norm(v)** to convince yourself that you must be careful. See the help page on **norm** for more information.

You can also compute the
angle between two vectors.

```
angle(v,w);
```
Recall that the angle θ between two vectors **v** and **w** satisfies
$$\mathbf{v} \cdot \mathbf{w} = \|\mathbf{u}\| \|\mathbf{v}\| \cos(\theta).$$

More on Subscripts and Sequences

Subscripts and sequences are intimately tied together. In Maple, any collection of objects separated by commas forms a *sequence*. Hence, a list is a sequence enclosed in square brackets, a set is a sequence enclosed in curly braces, and so on.

Sequences are generated in Maple with the **seq** command. The syntax of the **seq** command resembles that of the definite integral command, `int(f(x),x=1..5)`.

Here the seq *command*
generates an initial
segment of the sequence
of factorials.

```
F:=seq(i!,i=0..10);
```
$F[i]$ is now $(i-1)!$.

The seq *command will also*
support the s1, s2,...
form of subscripting if you
use a dot between the s
and the i.

```
seq(s.i,i=1..5);
```
The dot (period) is the *concatenation operator* in Maple. (To concatenate two strings of symbols means to put them end to end to form one; hence, **s.3** evaluates to **s3**.)

Like a loop, the seq *com-*
mand leaves the counter
with a value; you may wish
to unassign it.

```
i;
i:='i';
```
Some users prefer to unassign variables just as they finish with them. Others prefer to unassign them as the need arises.

You can use the seq *com-*
mand effectively to gen-
erate an argument for a
function of several vari-
ables, like augment.

```
S:=matrix(5,5);
seq(row(S,i),i=1..5);
T:=augment(");
```
This builds a new matrix having the rows of S as its columns. Hence, T is the transpose of S.

It is a good idea to unassign i.

```
i:='i';
```

You can also create the desired sequence manually. Which approach is more efficient depends on the number of terms.

```
row(S,1),
row(S,2),
row(S,3),
row(S,4),
row(S,5);
T:=augment(");
```

Notice the commas between the **row** commands.

Used at the right times, the **seq** command can save a lot of typing.

Polynomials and Matrices: The Cayley-Hamilton Theorem

There are a number of interesting ways in which polynomials and matrices interact. In this section you will see how Maple can be used to explore one of them.

You can compute the characteristic polynomial det($xI - A$) of a square matrix either directly or using Maple's charpoly *command.*

```
A:=matrix(3,3,(i,j)->i+j-1 mod 3);
p:=charpoly(A,x);
```

Note that the characteristic polynomial has degree 3 and constant term $(-1)^3 \det(A)$.

The Cayley-Hamilton Theorem, *named after its discoverers, says that every square matrix is a root of its own characteristic equation $C_A(x) = 0$.*

```
subs(x=A,p);
evalm(");
```

That certainly verifies the Cayley-Hamilton Theorem for the matrix A. But isn't it obvious that det($AI - A$) = 0? Yes, of course. But the Cayley-Hamilton Theorem makes the substitution $x = A$ after taking the determinant. In the first case, the answer is the number 0; in the second case, it is the matrix 0.

You can verify the Cayley-Hamilton Theorem in general for low dimensions using Maple's symbolic capabilities.

```
M:=matrix(3,3);
evalm(charpoly(M,M));
```

Perhaps some simplification would help.

You use the map *command to cause the specified command to work on the entries of the matrix.*

```
map(simplify,");
```
Most mathematical functions (sin, cos, etc.) are mapped onto the entries of a matrix by the **evalm** command. However, some Maple system commands, like **simplify**, are not; these are handled with the **map** command.

It follows from the Cayley-Hamilton Theorem that if p is any polynomial and if r is the remainder on dividing p by the characteristic polynomial of A, then $p(A) = r(A)$. Since the degree of the remainder is always smaller than the degree of the divisor, it follows that no polynomial in A need be written in a form with degree more than two.

Use the rem *command to compute the remainder on dividing one polynomial by another.*

```
rem(x^7-4*x^3+1,x^3-3*x^2-3*x+9,x);
```
Recall that given any two polynomials p and d, p can be written in the form

$$p = qd + r$$

where r is either 0 or has degree less than the degree of d.

The rem *command also works on polynomial functions.*

```
p:=x->x^7-4*x^3+1;
r:=x->rem(p(x),charpoly(A,x),x);
```

You can now verify that $p(A) = r(A)$.

```
p(A);
evalm(");
r(A);
evalm(");
```
You might prefer to verify that $p(A) - r(A) = 0$.

Polynomials and Matrices: Curve Fitting

Two points, (x_1, y_1) and (x_2, y_2), with distinct x-coordinates, determine a line and therefore a polynomial of the form $p(x) = ax + b$. Similarly, three points, (x_1, y_1), (x_2, y_2) and (x_3, y_3), determine a polynomial of the form $p(x) = a + bx + cx^2$, and so on.

Finding the equation of the polynomial

$$p(x) = a_0 + a_1 x + \ldots + a_n x^n$$

determined by $n + 1$ points in the form (x_1, y_1), (x_2, y_2), $\ldots, (x_{n+1}, y_{n+1})$, with distinct x-coordinates, requires solving a system of $n + 1$ equations in $n + 1$ unknowns. The system is generated by evaluating the equation $p(x_i) = y_i$, for $i = 1 \ldots n + 1$. In matrix form, the system is $M\mathbf{a} = \mathbf{y}$, where

$$
M = \begin{bmatrix}
1 & x_1 & x_1^2 & \cdots & x_1^n \\
1 & x_2 & x_2^2 & \cdots & x_2^n \\
 & & \vdots & & \\
1 & x_{n+1} & x_{n+1}^2 & \cdots & x_{n+1}^n
\end{bmatrix},
$$

$\mathbf{a} = (a_0, a_1, \ldots, a_n)$ and $\mathbf{y} = (y_1, y_2, \ldots, y_{n+1})$.

Hence, the coefficients of the polynomial $p(x)$ can be found by solving the matrix equation $M\mathbf{a} = \mathbf{y}$ for a. Note that $M = (x_i^{\,j-1})$. Matrices of this form are called *Vandermonde matrices*. It is a theorem that Vandermonde matrices are invertible, so the equation $M\mathbf{a} = \mathbf{y}$ has a unique solution.

In what follows, the polynomial

$$p(x) = a_0 + a_1 x + \ldots + a_4 x^4$$

passing through the points (1, 5), (2, 3), (3, 27), (4, 12), and (5, 2) is derived. Such data points might represent instrument readings taken at 1:00, 2:00, 3:00, 4:00, and 5:00 o'clock.

You can use Maple's vandermonde *command to generate the coefficient matrix M.*

```
y:=vector([5,3,27,12,2]);
M:=vandermonde([1,2,3,4,5]);
```
Note the form of M.

The vector of coefficients of p is the solution of the equation $M\mathbf{x} = \mathbf{y}$.

```
a:=linsolve(M,y);
```
The components of a are the coefficients of $p(x)$.

You can now compute the polynomial p either as an expression or as a function.

```
p:=x->dotprod(
a,vector([seq(x^i,i=0..4)]));
```
You might wish to double check the value of $p(x)$ at $x = 1, 2, 3, 4, 5$.

You can plot the graph of the polynomial p for more information.

```
plot(p(x),x=0..6);
```
As usual, you may have to experiment with the plot parameters to get the desired information.

Summary and Extensions

Maple knows all the standard matrix and vector functions: **det, transpose, inverse**, etc.

Every square matrix is a root of its own characteristic polynomial.

Vandermonde matrices arise naturally in curve fitting. It is a theorem that any Vandermonde matrix is invertible.

Maple's symbolic capabilities can be used to verify important theorems (with possibly obscure proofs) in low dimensions.

You can use the **seq** command to save typing in many cases.

Related Commands: companion.

Additional Activities

1. Compute the angle between the vectors $\mathbf{u} = (1, 2, 3, 5)$ and $\mathbf{v} = (3, -2, 2/3, 4)$. Convert the angle to degrees and get a floating point approximation.

2. Use the **vandermonde** and **linsolve** commands to find the polynomial $p = a_0 + a_1x + a_2x^2 + a_3x^3 + a_4x^4 + a_5x^5$ passing through the six points $(-6, 12)$, $(-3/2, -2)$, $(1/4, 3)$ $(3/4, 11)$, $(47, 33)$ and $(11, -27)$.

3. Let M be the matrix

$$\begin{bmatrix} 0 & 1 & 1 & 1 & 1 \\ 1 & 0 & 1 & 1 & 1 \\ 1 & 1 & 0 & 1 & 1 \\ 1 & 1 & 1 & 0 & 1 \\ 1 & 1 & 1 & 1 & 0 \end{bmatrix},$$

and let A be the submatrix obtained by deleting the fifth column of A. Compute the rank of A. Since the rank cannot exceed the number of columns, the matrix A is said to be of *full* rank. Show that the matrix $A^T A$ is invertible. It is a theorem that for any $m \times n$ matrix A of rank n, the matrix $A^T A$ is invertible. Can you see why?

4. It is a theorem that the matrix B obtained by interchanging the first and second rows of a square matrix A has determinant $-\det(A)$. Verify this for a symbolic 4×4 matrix A. Note: Since the determinant of a 4×4 matrix has 24 terms, you may prefer to verify that $\det(A) + \det(B) = 0$. (You may want to use the **swaprow** command.)

5. Let A be the matrix generated by the Maple command **A:=matrix(3,3, (i,j)->i+j-1 mod 3)**. Let

$$p(x) = x^{17} - 4x^{15} + 3/2x^6 - 1/2x^3 + 7x - 3,$$

and let $r(x)$ be the remainder on dividing $p(x)$ by the characteristic polynomial of A. Show that $p(A) = r(A)$.

6. Apply the **factor** command to the determinant of the 4×4 matrix **V := vandermonde([a,b,c,d])**. This will reveal one of the reasons why the Vandermonde matrix is so well known.

7. Every monic polynomial is the characteristic polynomial of a suitably chosen matrix. One of the simplest matrices having characteristic polynomial p is called the *companion matrix* of p and denoted **companion(p,x)** in Maple.

a. Verify that $p = x^{12} - 32x^{10} + 11x^4 - 32x + 17$ is the characteristic polynomial of its companion matrix.

b. Find a formula for the companion matrix.

4.4 Maple's Basis and Dimension Commands

One of the most useful ways to describe a vector space is by specifying a basis; Maple has a number of built-in commands which can assist you in doing this.

As always...

```
with(linalg);
alias(I=&*());
```

If M is any matrix, the nonzero rows of the reduced row echelon form are a basis for the row space of M.

```
M:=matrix(9,7,(i,j)->i+j mod 2);
F:=rref(M);
B:={row(F,1),row(F,2)};
```

The set *B* of nonzero rows of *F* is a basis for the row space of *M*.

Maple's rowspace *command will compute this basis for you.*

```
rowspace(M);
```

Hence, the row space of *M* consists of all vectors of the form (x, y, x, y, x, y, x).

The colspace *command works analogously.*

```
colspace(M);
```

The colspace command applies **rowspace** to the transpose of the specified matrix. The vectors returned are a *basis* for the column space of *M*. Note that the column space consists of all vectors of the form (x, y, x, y, x).

You can also use Maple's basis *command to obtain a basis from any spanning set.*

```
for i to 9
  do
    u.i:=vector(7,j->i+j)
  od;
SpanList:=[seq(u.i,i=1..9)];
B:=basis(SpanList);
```

The argument to the **basis** command can be any set or list of vectors. The result is a subset or sublist of the vectors passed to **basis**.

If you require an orthogonal basis, you can apply Maple's GramSchmidt *command.*

```
GS:=GramSchmidt(B);
```

The argument to **GramSchmidt** can be either a set or a list of vectors. The result is an orthogonal set or list of vectors with the same span as the original. The vectors passed to **GramSchmidt** need not be linearly independent, but, if they are not, **0** will appear among the vectors returned. To see how this works, you might wish to apply **GramSchmidt** to **SpanList**.

The (nonzero) vectors can be normalized by dividing each by its norm.

```
T:=[seq(GS[i]/norm(GS[i],2),i=1..2)];
```
The ith element of the list GS is $GS[i]$. Recall that the standard vector norm is denoted **norm(v,2)** in Maple.

The range *command returns a basis for the range of the matrix function* $\mathbf{x} \to M\mathbf{x}$.

```
range(M);
```
The **range** command is a synonym for the **colspace** command; the two can be used interchangeably.

You can also obtain the dimension of the row or column space without computing a basis.

```
rank(M);
```
You might wish to compare the rank of M to the rank of M^T. Your text says they should agree. You might also wish to compare the rank of M to the rank of the product $M^T M$. Can you think of any reason why they should agree?

You can use the nullspace *command to get a basis for the kernel of the command* $\mathbf{x} \to M\mathbf{x}$.

```
K:=nullspace(M);
```
The **nullspace** command is also called **kernel**.

You may wish to compare the output of nullspace *to that of* linsolve.

```
linsolve(M,vector(9,i->0));
```
The **linsolve** routine returns a general, parameterized solution, from which it is easy to derive a basis for the solution space; **nullspace** returns a basis directly.

Sums and Intersections of Subspaces

Assume U and V are subspaces of R^n *spanned by* $C = \{\mathbf{u}_1, \mathbf{u}_2, \ldots, \mathbf{u}_r\}$ *and* $D = \{\mathbf{v}_1, \mathbf{v}_2, \ldots, \mathbf{v}_s\}$, *respectively.*

```
C:={seq(vector(7,j->min(i,j)),i=1..5)};
D:={seq(vector(7,j->max(i,j)),i=1..4)};
```

In this particular case, both C and D are linearly independent sets.

```
basis(C);
basis(D);
```
Hence U and V have dimensions 5 and 4, respectively.

You can use Maple's sumbasis command to find a basis for the subspace $U + V$.

```
sumbasis(C,D);
```
The subspace $U + V$ is spanned by $C \cup D$. Notice that $U + V$ has dimension 6.

You can also use Maple's intbasis command to find a basis for the intersection $U \cap V$.

```
intbasis(C,D);
```
Note the relation between $\dim(U+V)$ and $\dim(U)+\dim(V) - \dim(U \cap V)$. Can you prove this in general?

Summary and Extensions

Maple provides the commands **rowspace, rowspan, colspace, colspan, range, kernel, nullspace,** and **basis** for finding bases of subspaces.

You can use the **GramSchmidt** command to get an orthogonal basis for a subspace.

Related commands: rowspan, colspan.

Additional Activities

1. Find a basis B for the subspace U of R^5 spanned by the vectors $\mathbf{v}_1 = (1,3,5,7,9)$, $\mathbf{v}_2 = (3,5,7,9,11)$, $\mathbf{v}_3 = (5,7,9,11,13)$, $\mathbf{v}_4 = (7,9,11,13,15)$.

2. Let M be the augmented matrix $M = [\mathbf{v}_1, \mathbf{v}_2, \mathbf{v}_3, \mathbf{v}_4]$ with ith column vector $\mathbf{v}_i$ from Activity 1. Find bases for the nullspace and range of M, and find the rank of M.

3. Use **GramSchmidt** on the basis B of Activity 1 to find an orthogonal basis for U.

4. Use the result of activity 3 to find an orthonormal basis T for the subspace U of Activity 1.

5. Use **randmatrix** with the **sparse** option to generate a 10×10 matrix M. Use Maple's **intbasis** command to verify that the rowspace of M has a trivial intersection with the nullspace of M.

6. Pick two subsets B and C of R^{10} at random. (You could use the **randmatrix** command and use the rows of the result.) Use Maple's **sumbasis**, **basis**, and **intbasis** commands to verify that the dimension of the span of $B \cup C$ is the dimension of the span of B plus the dimension of the span of C minus the dimension of the intersection of the span of B and the span of C.

4.5 Linear Transformations

You can define a linear transformation using either an arrow-style definition or a **proc** definition.

This defines a linear transformation from R^5 to R^3.

```
L:=x->vector([2*x[1]-x[2],
x[4]+3*x[5],3*x[5]+1/2*x[4]]);
```
Note that the Maple definition of L does not restrict its domain to R^5—or even to vectors for that matter.

As defined, L cannot be applied to expressions—even if they evaluate to a vector in R^5.

```
u:=vector(5,i->i);
v:=vector(5,i->i!);
L(u+v);
```
You could use **L(evalm(u+v))** here.

Using a proc-style definition, you can define linear transformations that can be applied to expressions. The linear transformation T defined here has this property.

```
T:=
proc(x)
   local y;
   y:=evalm(x);
   vector([2*y[1]+3*y[2],
   -3*y[1]+y[3],3*y[1]-2*y[3]]);
end;
```
Notice that T applies **evalm** to its own argument. The second line of the definition confines the effect of assigning y to the internal workings of the procedure itself.

The linear transformation T can be applied to expressions involving both numerical and symbolic vectors.

```
u:=vector([5,-7,9]);
v:=vector(3);
T(u+k*v);
```

The standard matrix of a linear transformation can be easily determined by applying it to a symbolic vector.

```
T(v);
A:=matrix([
[2,3,0],
[-3,0,1],
[3,0,-2]]);
```

The (i,j)-entry of A is the coefficient of **v**[j] in the ith component of $T(\mathbf{v})$.

You can also use Maple's genmatrix *command to have Maple compute A for you.*

```
genmatrix(
[T(v)[1],T(v)[2],T(v)[3]],
[v[1],v[2],v[3]]);
```

Note that the first argument is the list of components of $T(\mathbf{v})$ and the second argument is the list of components of **v**.

You can make the calculation of the matrix even easier by converting the vectors to lists.

```
TT:=convert(T(v),list);
vv:=convert(v,list);
genmatrix(TT,vv);
```

This is particularly helpful for larger matrices.

You can easily verify that $T(\mathbf{v}) = A\mathbf{v}$ in general.

```
T(v);
evalm(A&*v);
```

You can find the kernel and range of T from A.

```
kernel(A);
range(A);
```

If a linear transformation L is specified by a collection of equations $L(\mathbf{b}_i) = \mathbf{d}_i$, $i = 1 \ldots k$, where $\{\mathbf{b}_1, \mathbf{b}_2, \ldots, \mathbf{b}_k\}$ is a basis for V, you can easily create a Maple procedure which returns $L(\mathbf{v})$ for any vector **v** in V. Note that if $\mathbf{v} = a_1\mathbf{b}_1 + \ldots + a_5\mathbf{b}_5$ then $L(\mathbf{v}) = a_1\mathbf{d}_1 + \ldots + a_5\mathbf{d}_5$.

Here are two collections $\mathbf{b}_1,\ldots,\mathbf{b}_5$ and $\mathbf{d}_1,\ldots,\mathbf{d}_5$ of vectors in R^7. The vectors $\mathbf{b}_1,\ldots,\mathbf{b}_5$ are a basis for the subspace V they span.

```
for j to 5
  do
    b.j:=vector(7,i->min(i,j))
  od;
for j to 5
  do
    d.j:=vector(7,i->
    sum(r,r=max(j-i+1,0)..j))
  od;
```

If $\mathbf{v} = a_1\mathbf{b}_1+\ldots+a_5\mathbf{b}_5$ is any vector in V, then the coefficients $a_1, a_2, \ldots, a_5$ can be obtained by using linsolve *with the augmented matrix $B = [\mathbf{b}_1,\ldots,\mathbf{b}_5]$.*

```
B:=augment(seq(b.i,i=1..5));
v:=vector([1,7,9,4,2,2,2]);
a:=linsolve(B,v);
```

You may recall that if a vector **u** lies outside of the span of $[\mathbf{b}_1,\ldots,\mathbf{b}_5]$ then **linsolve(B,u)** will give no response, not even an error message. The vector $\mathbf{a} = (a_1, a_2, \ldots, a_5)$ is called the coordinate vector of **v** with respect to the basis $B = \{\mathbf{b}_1,\ldots,\mathbf{b}_5\}$.

$L(\mathbf{v})$ is now given by $a_1\mathbf{d}_1+\ldots+a_5\mathbf{d}_5 = D\mathbf{a}$, where D is the augmented matrix $[\mathbf{d}_1,\ldots,\mathbf{d}_5]$.

```
D:=augment(seq(d.i,i=1..5));
w:=evalm(D&*a);
```

Notice that the result is assigned the name **w**.

These steps can easily be combined into a procedure.

```
L:=x->evalm(D&*linsolve(B,x));
```

Note that L will apply the same steps to any vector that were applied to **v**. You might wish to verify this by comparing $L(\mathbf{v})$ with **w**.

Orthogonal Projection

If U is a subspace of R^n and $\mathbf{v} \in R^n$, then the element of U closest to **v** is the orthogonal projection of **v** onto U. The orthogonal projection of a vector **v** in R^n onto the column space of a matrix A is the vector **w** closest to **v** for which the equation $A\mathbf{x} = \mathbf{w}$ has solution. The vector **w** is unique and the map $\mathbf{v} \rightarrow \mathbf{w}$ is a linear transformation. The solution of the equation $A\mathbf{x} = \mathbf{w}$ is called the least squares (or "best approximate") solution of the equation $A\mathbf{x} = \mathbf{v}$. In Maple, the general best approximate solution of the equation $A\mathbf{x} = \mathbf{v}$ is denoted **leastsqrs(A,v)**.

You can use Maple's `leastsqrs` *command to get an approximate solution of a vector equation* $A\mathbf{x} = \mathbf{v}$. *If an exact solution exists, it will be given.*

```
R:=randmatrix(5,4);
v:=vector([1,1,1,1,1]);
a:=leastsqrs(R,v);
```
The vector **a** is chosen so that the distance $R\mathbf{a} - \mathbf{v}$ is minimized.

You can use the result **a** *of* `leastsqrs(R,v)` *to compute the orthogonal projection of* **v** *onto the column space of R.*

```
w:=evalm(R&*a);
```
The vector **w** is the best approximation to **v** that lies in the column space of the matrix A. The vector **w** can also be obtained as the sum

$$(\mathbf{v} \cdot \mathbf{r}_1)\mathbf{r}_1 + (\mathbf{v} \cdot \mathbf{r}_2)\mathbf{r}_2 + \ldots + (\mathbf{v} \cdot \mathbf{r}_5)\mathbf{r}_5$$

where $\{\mathbf{r}_1, \mathbf{r}_2, \ldots, \mathbf{r}_5\}$ is an orthonormal basis for the column space of R.

Summary

You can easily write a Maple procedure to implement the definition of a linear transformation as a function.

Maple's symbolic capabilities allow you to find the standard matrix of a linear transformation from R^n to R^m.

You can automate the computation of the matrix of a linear transformation by using **genmatrix**.

You can convert vectors to lists (or sets) using Maple's **convert** command.

Use **leastsqrs** to get an approximate solution of an equation $A\mathbf{x} = \mathbf{b}$. The exact solution is returned if one exists.

You can use **leastsqrs** and matrix multiplication to compute the orthogonal projection of a vector onto the column space of a matrix.

Additional Activities

1. Let $A = (a_{ij})$ be the 7×7 matrix defined by
    $$a_{ij} = \begin{cases} 0 & \text{if } i = j \\ 1 & \text{if } i \neq j \end{cases}.$$
    For each $i = 1 \ldots 7$, let $\mathbf{b}_i$ be the ith row vector of A and let $\mathbf{c}_i = (i, i+1, i+2, i+3, i+4, i+5, i+6)$. Assume that T is a linear transformation satisfying $T(\mathbf{b}_i) = \mathbf{c}_i$, for $i = 1 \ldots 7$. Find a formula for T and the standard matrix of T.

2. Find the kernel and range of the linear transformation T of Activity 1.

3. Given a basis $B = \{\mathbf{b}_1, \mathbf{b}_2, \ldots, \mathbf{b}_k\}$ for a vector space V, any vector $\mathbf{v}$ in V has unique expression
    $$\mathbf{v} = a_1\mathbf{b}_1 + a_2\mathbf{b}_2 + \ldots + a_k\mathbf{b}_k$$
    The k-tuple $(a_1, a_2, \ldots, a_k)$ is called the B-coordinate vector of $\mathbf{v}$. The map which takes each vector $\mathbf{v}$ to its B-coordinate vector is a linear transformation. Let $B = \{\mathbf{b}_1, \ldots, \mathbf{b}_6\}$ be the basis for R^6 defined by $\mathbf{b}_i[j] = \min(i, j)$. Define a procedure **Coord** which takes every vector $\mathbf{v}$ in R^6 to its B-coordinate vector.

4. Find the standard matrix of the linear transformation **Coord** of Activity 3.

5. If B and D are two bases for a vector space V, then the map which takes the B-coordinate vector of every vector $\mathbf{v}$ to the D-coordinate vector of $\mathbf{v}$ is a linear transformation. Let B be as in Activity 3, and let D be the basis for R^6 defined by $d_j[i] := \text{sum}(r, r = \max(j - i + 1, 0) \ldots j))$. Define a function L which takes the B-coordinate vector of any $\mathbf{v}$ in R^6 to the D-coordinate vector of $\mathbf{v}$.

6. Find the standard matrix of the linear transformation L of Activity 5.

7. Find the "best approximate solution" to the equation $Mx = $ **v**, where **v**$= (1, 2, 3, 4, 5)$ and $M = (m_{ij})$ is the 5×3 matrix with $m_{ij} = 1$, for all i, j.

8. In Activity 7, find the orthogonal projection of the vector **v**$= (1, 2, 3, 4, 5)$ onto the column space of the matrix M.

4.6 Eigenvalues and Eigenvectors

Eigenvalues and eigenvectors, also known as characteristic values and characteristic vectors, are among the most frequently applied topics commonly studied in undergraduate mathematics. For example, if multiplication by A represents the action of the forces on a physical system with components represented by the vector **v**, and if $A\mathbf{v} = \lambda\mathbf{v}$, then the components of the system are all changing at the same rate. If $\lambda = 1$, then the system is static.

In this section, all matrices have rational entries. This implies that the characteristic polynomials of the matrices considered will have rational coefficients, but it does not imply that the eigenvalues are rational. Maple is capable of describing the eigenvalues and eigenvectors of any rational matrix—within the limits imposed by the time and resources available to it.

If you are beginning a new session:

```
with(linalg);
alias(I=&*());
```

Here is a 4 × 4 matrix which is fairly typical of those commonly used to study eigenvalue/eigenvector problems,

```
f:=
  proc(i,j)
    if i=j
      then i+j-1
      else i+j+1
    fi
  end;
A:=matrix(4,4,f);
```

Maple has built-in commands for computing the characteristic matrix and characteristic polynomial.

```
charmat(A,x);
charpoly(A,x);
```
Note that a second argument has to be specified in each case. The second argument can be either an undefined variable, as it is here, or a constant.

Use the `charpoly` *and* `solve` *commands to find the eigenvalues.*

```
solve(charpoly(A,x)=0,x);
```
The eigenvalues are the terms of the sequence returned.

Alternately, you can use the built-in `eigenvals` *command to perform the same operations.*

```
eigenvals(A);
lambda:={"};
```
Note that there are three distinct eigenvalues λ_1, λ_2, and λ_3, two of multiplicity one and one of multiplicity two; placing them in a set removes the duplication.

For each i, the λ_i-eigenspace is the nullspace of the matrix $\lambda_i I - A$.

```
Z:=vector(4,i->0);
linsolve(charmat(A,lambda[1]),Z);
```
This is a parametric description of the vectors in the λ_1-eigenspace.

You can use the `null-space` *command to get a basis for the eigenspaces.*

```
nullspace(charmat(A,lambda[1]));
```
The set returned is a basis for the λ_1-eigenspace. You may wish to repeat the process for the other two eigenvalues.

One special property of the matrix A is that its characteristic polynomial has no factors of degree greater than two.

```
factor(charpoly(A,x));
```
The factor command returns a factorization of the form $r p_1^{e_1} p_2^{e_2} \ldots p_k^{e_k}$, where r is a rational number and each p_i is a nonconstant polynomial with integer coefficients. The approach used for A will work for any (rational) matrix, as long as none of the factors p_i of the characteristic polynomial has degree three or greater.

In general, however, characteristic polynomials can have factors of any degree.

```
p:=x^6-12*x^5+30*x^4+32*x^3-31*x^2-4*x+3;
B:=companion(p,x);
factor(charpoly(B,x));
```
The matrix B is called the *companion matrix* of the polynomial p. Its characteristic polynomial has two factors of degree three.

In this case, the descriptions of the eigenvalues are too complicated for the matrix equation $(\lambda I - B)\mathbf{x} = \mathbf{0}$ to be solved by either nullspace *or* linsolve.

```
eigenvals(B);
lambda:=";
nullspace(charmat(B,lambda[1]));
```
The answer is obviously incorrect; an eigenvalue cannot have a trivial eigenspace.

The complexity of the description by radicals of a root of a cubic or quartic factor is a barrier which complicates the computation of the eigenvectors of matrices larger than 2×2. To break through this barrier, Maple provides a simpler representation of the roots of the characteristic polynomial that will function with **nullspace** and **linsolve**: If p is a polynomial in x, then **RootOf(p,x)** is a parameter which can stand for any of the roots of the polynomial p.

The RootOf *command is typically applied to the factors of the characteristic polynomial. Note that a factor of degree n has exactly n (distinct, but possibly complex) roots.*

```
facts:=factor(charpoly(B,x));
p1:=op(facts)[1];
lambda[1]:=RootOf(p1,x);
```
The **op** command returns the sequence of factors. In this case, $r = 1$ in the factorization $rp_1^{e_1}p_2^{e_2}\ldots p_k^{e_k}$ of the characteristic polynomial, so Maple simplifies the factorization to $p_1^{e_1}p_2^{e_2}\ldots p_k^{e_k}$. Hence, **op(facts)[1]** is p_1. Each root of each p_i has multiplicity e_i as a root of p.

Either nullspace *or* linsolve *can now be used to obtain a description of the eigenvectors.*

```
nullspace(charmat(B,lambda[1]));
```
It is important to keep in mind that the (apparently) single basis returned actually represents three different bases, one for each of the three roots of p_1.

You can simplify the notation considerably by aliasing an unassigned variable to the RootOf.

```
alias(t=lambda[1]);
NS:=nullspace(charmat(B,t));
```
Notice that the result has been assigned the name NS.

You can use Maple's allvalues *command to convert a* RootOf *into radical form whenever this is possible.*

```
all:=allvalues(t);
```
The translation to radical form can be made as long as the parameter is a **RootOf** of a polynomial of degree four or less. There is no description by radicals, in general, of the roots of a polynomial of degree five or greater. When

allvalues cannot find a description by radicals, it will attempt to find decimal approximations of the exact answers.

If you wish, you can substitute the radical forms for t in the basis vector(s).

```
ES[1]:=subs(t=all[1],NS);
```
Similar statements substituting t=all[2] and t=all[3] into NS give the bases for the eigenspaces associated with the other two roots of p_1.

A few properties of RootOf *worth noting:*

```
RootOf(3,x);
RootOf(2*x-3/2,x);
RootOf((x^2-1)^2,x);
```
In the first case, there are no roots. In the second case, the root is computed and returned. In the third case, some simplification is performed.

If you want all of the eigenvalues of a matrix in RootOf *form, you can use the* eigenvals *command with the* implicit *option.*

```
eigenvals(A,implicit);
```
You use the implicit option by using "implicit" as a second argument to eigenvals.

Once you are familiar with the use of **RootOf**, you can interpret the result returned by Maple's **eigenvects** command.

You can use eigenvects *to get both the eigenvalues and eigenvectors.*

```
eigenvects(A);
```
Note that the result is a sequence of lists. Each list is of the form

$$[\lambda, n, \{\mathbf{v}_1, \mathbf{v}_2, \ldots, \mathbf{v}_s\}]$$

where λ is an eigenvalue (either a rational number or a **RootOf**), n is the multiplicity of λ as a root of the characteristic polynomial, and the set $\{\mathbf{v}_1, \mathbf{v}_2, \ldots, \mathbf{v}_s\}$ is a basis for the λ-eigenspace.

It is frequently important to know if the eigenvalues of a matrix are rational, real, or complex. If the eigenvalues are in radical form, Maple's complex number evaluator, **evalc**, will attempt to write them in the form $a + bi$ with a and b real.

Use evalc *to see if the eigenvalues are real.*

```
allc:=evalc([all]);
```
If applied to a list or set, **evalc** is automatically mapped onto the entries of the list or set. This is true for lists as well, but not for sequences or arrays.

Recall that **all** was assigned the value returned by **all-values(t)** above.

Ah, the complex terms are gone; all the eigenvalues are real.

If you want more information on the eigenvalues, you can also apply the **evalf** command. However, there is always some possibility that rounding errors may mislead you.

Use evalf *for a floating point evaluation.*

```
evalf(");
```
If applied to a set, **evalf**, like **evalc**, is automatically mapped onto the elements of the set.

You might like to try using the **evalf** and **evalc** commands in the other order to see a small effect of rounding errors.

For factors of degree five or greater, floating point approximation and graphing are the only tools available other than **RootOf**.

Techniques for exploring roots of polynomials with Maple were discussed at some length in Chapter 2.

You unalias *a variable by* aliasing *it to itself.*

```
alias(t=t);
```
Notice that whenever the **alias** command is used, it lists all aliased variables.

Summary

You can use either the **factor** command or the **eigenvals** command to investigate the eigenvalues of a matrix.

The **RootOf** command makes it possible to describe the eigenvalues and eigenvectors of any rational matrix.

Use either **linsolve** or **nullspace** to find the eigenvectors associated with a particular eigenvalue.

Use the **allvalues** command to convert **RootOf** to radical expressions if possible. If exact radical answers are not possible, floating point approximations will be returned.

You can use **evalc** to determine if eigenvalues are real or complex. You can also use **evalf** for the same purpose, though it is often better to use **evalc** first.

Additional Activities

1. Find all the eigenvalues and eigenvectors of the 9×9 matrix $A = (a_{ij})$ with $a_{ij} = 1$ for all i and j.

2. Find all the eigenvalues and eigenvectors of the companion matrix of the polynomial

    $$p = 1 - x + x^2 - x^3 + x^4$$

    Determine which of the eigenvalues are rational, which are real, and which are complex.

3. Find the eigenvalues and eigenvectors of the 4×4 matrix $B = (b_{ij})$ where, for each i and j, b_{ij} is the smaller of i and j. Determine which of the eigenvalues are rational, which are real, and which are complex.

4. Find the eigenvalues and eigenvectors of the 3×3 matrix $Q = (q_{ij})$ where, for each i and j, $q_{ij} = i/j$. Determine which of the eigenvalues are rational, which are real, and which are complex.

4.7 Diagonalization and Similarity

There are two problems associated with diagonalization. The first is to determine if a given matrix A is diagonalizable (i.e., similar to a diagonal matrix). If this is the case, then you may also want to find a diagonalizing matrix P (i.e., an invertible matrix P satisfying $P^{-1}AP = D$, where D is a diagonal matrix).

Recall that an $n \times n$ matrix A is diagonalizable if, and only if, it has n linearly independent eigenvectors. This, in turn, is equivalent to the condition that for each eigenvalue λ of A, the dimension of the λ-eigenspace is equal to the multiplicity of λ as a root of the characteristic polynomial. The restriction to rational matrices continues.

Diagonalization

If you have not already done so:

```
with(linalg);
alias(I=&*());
```

The matrix A is a good, simple first example.

```
f:=
  proc(i,j)
    if i=j
      then 0
      else 1
    fi
  end;
A:=matrix(9,9,f);
```

The matrix A is symmetric, and therefore diagonalizable. The diagonalizability is also obvious from the information returned by eigenvects.

```
eigsys:=eigenvects(A);
```
The basis for the ith eigenspace is the third component of **eigsys[i]**. Notice that for each eigenvalue λ, the dimension of the λ-eigenspace is the same as the multiplicity of λ as a root of the characteristic polynomial.

The basis vectors of the eigenspaces are used to form a diagonalizing matrix.

```
ES[1]:=eigsys[1][3];
ES[2]:=eigsys[2][3];
```

You can use op *and* augment *to construct the diagonalizing matrix.*

```
P:=augment(op(ES[1]),op(ES[2]));
```
The **op** command returns the contents of the sets as sequences Two sequences separated by a comma form a sequence, so this command applies **augment** to the sequence of eigenvectors in the bases.

Since there are no horrendously complicated expressions in either A or P, Maple can verify that $P^{-1}AP$ is diagonal.

```
evalm(inverse(P)&*A&*P);
```
The eigenvalues appear on the diagonal according to the ordering of the vectors in *P*. You might wish to compare this result with that obtained if the roles of *ES*[1] and *ES*[2] are reversed in the definition of *P*.

The matrix B defined here presents an obstacle not encountered in diagonalizing A.

```
g:=
  proc(i,j)
    if i=j
      then i+j-1
      else i+j
    fi
  end;
B:=matrix(5,5,g);
```

As a rule, you will probably want to apply `eigenvects`.

```
eigsys:=eigenvects(B);
```
It is clear that there are no complex components in any of the eigenspace basis vectors associated with the eigenvalue -1. The situation regarding the other eigenvectors is somewhat less clear.

To simplify the descriptions returned by `eigenvects`, *it is helpful to give an* `alias` *to any* `RootOf` *which appears in an eigenvector.*

```
alias(t=RootOf(x^2-28*x-79,x));
```
If your system returns a different RootOf, you use it instead of this one. If your system assigned a label (e.g., %1) to the **RootOf**, you can alias the variable to the label instead; the effect is the same.

The computation of the λ-eigenspace parallels the computations of the eigenspaces of A for $\lambda = -1$.

```
ES[1]:=eigsys[1][3];
```
One of the eigenvalues is in **RootOf** form; in our case the second one. If the order of the results returned by eigenvects is different in your case, you will need to adjust the subscripts accordingly.

It is likely you will want to convert the `RootOf`s *to radical form for describing the diagonalizing matrix.*

```
all:=allvalues(t);
```
The conversion is not possible in all cases, but it is possible, and probably desirable, in this case.

You can obtain the other two eigenspace bases by simply using subs.

```
ES[2]:=subs(t=all[1],eigsys[2][3]);
ES[3]:=subs(t=all[2],eigsys[2][3]);
```

You can now build the diagonalizing matrix in the same manner you used for A.

```
Q:=augment(op(ES[1]),op(ES[2]),
op(ES[3]));
```

For matrices with several eigenspaces, you may want to use the seq *command to save some typing.*

```
augment(seq(op(ES[i]),i=1..3));
```
The definition of "several" is personal; this is shorter than the style used to define Q.

The seq *command leaves the subscript with a value— you will probably want to unassign it to avoid future problems.*

```
i;
i:='i';
```

Here is an example of a matrix with a characteristic polynomial with an irreducible quartic factor.

```
p:=sum(x^i,i=0..4);
C:=companion(p,x);
```
The characteristic matrix of C is p.

The eigenvects *command will give you the information you need. This time there are no rational eigenvalues.*

```
eigsys:=eigenvects(C);
```
The result is a single list. While it is arguably reasonable to allow the list to be addressed as **eigsys[1]**, in fact this would return the first term of the list. In this case, the list itself is assigned the name **eigsys**.

You can create a formal description of a diagonalizing matrix R by simply substituting variable names for the RootOf. *In our case, the* RootOf *is labeled as %1.*

```
r:=%1;
for i to 4
  do
    ES[i]:=subs(r=r.i,eigsys[3])
  od;
```
By assigning **r** to %1, you attach your own permanent label to the **RootOf**. (The labels %1, etc., are attached depending on the values of the system variables **labeling** and **labelwidth**. The author's system has **label-**

ing=true and labelwidth=20. For more information, use ?interface.)

A formal description of a diagonalizing matrix can now be built.

```
R:=augment(seq(op(ES[i]),i=1..4));
```
This is a meaningful description with the addendum that r_1, r_2, r_3, r_4 are the four roots of the polynomial $p = 1 + x + x^2 + x^3 + x^4$.

Since the polynomial involved is of degree four, it is possible to describe the roots by radicals.

```
all:=allvalues(r);
```
Surprisingly, in this case, the descriptions are not too complicated.

Compare the formal description to the result of using the radical forms in the matrix.

```
for i to 4
   do
      FS[i]:=subs(r=all[i],eigsys[3])
   od;
RR:=augment(seq(op(FS[i]),i=1..4));
```
It could be argued that R is simpler to understand than RR. Moreover, for factors of degree five or greater, the option of using **allvalues** may not be available.

Similarity and Smith Form

Two (square) matrices A and B are *similar* if B is of the form $P^{-1}AP$, and—surprisingly—this is the case if, and only if, their characteristic matrices $xI - A$ and $xI - B$ are *equivalent* (i.e., if, and only if, $xI - B$ can be obtained from $xI - A$ by a sequence of elementary row and column operations). Every matrix M with polynomial entries can be transformed into an equivalent diagonal matrix $S = \text{diag}(1, \ldots, 1, p_1, p_2, \ldots, p_k, 0, \ldots, 0)$ where, for each i, p_i is a nonconstant factor of p_{i+1}. The matrix S is uniquely determined by M and is called the *Smith Form* of M. Two matrices are equivalent if, and only if, they have the same Smith form. Hence, two numerical matrices A and B are similar if, and only if, their characteristic matrices $xI - A$ and $xI - B$ have the same Smith Form. (The elementary operations **mulrow** and **mulcol** are restricted to multiplication by scalars to make the operations reversible. Notice that this restriction is not necessary for **addrow** and **addcol**.)

Maple can compute the Smith form of any matrix with polynomial entries.

```
M:=matrix(3,3,(i,j)->x^i-j);
smith(M,x);
```

Smith form can be used to determine if two matrices are similar. (Although the current assumption is that all matrices have rational entries, the theory is not restricted to this case.)

```
d:=
  proc(i,j)
    if i<j
      then 0
      else 1
    fi
  end;
G:=matrix(4,4,d);
H:=transpose(G);
smith(charmat(G,x),x);
smith(charmat(H,x),x);
```

In this case, the Smith forms of the characteristic matrices are the same, so the matrices are similar. (Actually, every matrix is similar to its transpose.)

Similarity and Frobenius Form

It follows from the previous section on Smith form that the polynomials on the diagonal of the Smith form of the matrix $xI - A$ determine to what matrices A is similar. Not surprisingly, the companion matrices of these polynomials also determine the "similarity class" of A. The block diagonal matrix F (with blocks that are the companion matrices of these nonconstant polynomials) is called the Frobenius form of A.

There are two clear advantages of the Froebenius form. One is that the Frobenius form of a rational matrix is rational. Another is that a matrix is similar to its Frobenius form. You can obtain the Frobenius form (sometimes called the rational canonical form, though this term is not used uniformly) with the **frobenius** command.

If you have not already done so:	```with(linalg);``` ```alias(Id=&*());```

Consider the matrix G defined here.

```
d:=
  proc(i,j)
    if i<j
      then 0
      else 1
    fi
  end;
G:=matrix(4,4,d);
```

G is the same matrix used in the previous section.

Compute the Frobenius form of G.

```
FG:=frobenius(G);
```

In this case, the Frobenius form is the companion matrix of the characteristic polynomial.

Compare the Smith forms of $xI-G$ and $xI-FG$.

```
smith(charmat(G,x),x);
smith(charmat(FG,x),x);
```

Note that they are the same, although you may have to expand the (4,4)-entry of the first matrix to see this. It follows that G is similar to FG.

Summary

Using **eigenvects**, it is easy to determine if a (rational) matrix is diagonalizable.

If a matrix A is diagonalizable, you can build a diagonalizing matrix P from the eigenspace bases returned by **eigenvects** by using **op** and **augment**.

You can determine if two matrices A and B are similar by comparing the Smith forms of their characteristic matrices.

You can determine if two matrices A and B are similar by comparing their Frobenius forms.

Additional Activities

1. Let $A = (a_{ij})$ be the 5×5 matrix with

    $$a_{ij} = \begin{cases} i + j - 1 & \text{if } i = j \\ i + j + 1 & \text{if } i \neq j \end{cases}.$$

    Determine if A is diagonalizable and, if so, find a diagonalizing matrix P.

2. Let $B = (b_{ij})$ be the 5×5 matrix with

    $$b_{ij} = \begin{cases} 2 & \text{if } i = j \\ 1 & \text{if } i \neq j \end{cases}.$$

    Determine if A is diagonalizable and, if so, find a diagonalizing matrix P.

3. Let $H = (h_{ij})$ be the 4×4 matrix defined by $h_{ij} = 1/(i + j - 1)$. Determine if H is diagonalizable and, if so, find a diagonalizing matrix. (H is the 4×4 *Hilbert matrix*.)

4. Let $p = 1 - x + x^2 - x^3 + x^4$ and $q = 1 - x + x^2$ have companion matrices A_{11} and A_{22}, respectively. Let A be the 6×6 matrix described in block form as

    $$A = \begin{bmatrix} A_{11} & 0 \\ 0 & A_{22} \end{bmatrix}.$$

    Let C be the companion matrix of pq. Use the Smith forms of the characteristic matrices of A and C to show that A and C are similar.

5. Use the Frobenius forms of the matrices A and C of Activity 4 to show they are similar.

6. For three randomly generated 4×4 matrices A, show that A and its Frobenius form F are similar by verifying that $xI - A$ and $xI - F$ have the same Smith form.

5 Differential Equations

This chapter exploits Maple's powerful graphical, numerical, and symbolic capabilities to aid in understanding the often complex nature of differential equations.

5.1 Introduction

Maple requires some setup to take advantage of the routines described in this chapter.

The **ODE.m** file, which contains a number of routines you will use to analyze differential equations, is not part of Maple. This file must be read into each session in which it is used. Once this file is in the directory or folder which contains Maple, you can read it into your Maple session. You may need to create the file called **ODE.m**. The following three commands will create the file.

Making the ODE *file accessible to a Maple V session.*

```
with(share);
read ``.sharename.`/plots/ODE`;
save `ODE.m`;
```
Notice the direction of all the single backquote marks (`). Exit Help after entering the **with(share)** statement. The **read** statement brings the needed commands into the Maple session where they are saved by the **save** statement. To make sure this procedure has worked, you should exit Maple and then open the Maple application again.

Access the ODE.m *file in*
each new Maple session.

```
read 'ODE.m';
```

This statement should be entered in any session where you wish to use the **ODE.m** file. Maple's capabilities along with the procedures contained in the **ODE** file can be used to enhance understanding of the structure of differential equations. Maple can find explicit solutions to many differential equations. In cases where no closed form solution exists, it can be used to implement numerical schemes to approximate solutions.

First Order Differential Equations

You first consider differential equations which can be written in the following form

$$\frac{dy}{dt} = f(t, y). \tag{1}$$

A solution is a continuous function of t, which when substituted for y, satisfies equation(1). You first need to translate the differential equation into a format Maple can recognize. The derivative of **y** with respect to **t** is represented in Maple by **diff(y(t),t)**. Consider solving the differential equation

$$\frac{dy}{dt} = t + y.$$

Enter the differential equa-
tion.

```
deq1:= diff(y(t),t) = t + y(t);
```

There is an alternative to using **diff(y(t),t)** to represent the derivative of **y** with respect to **t** which is **D(y)(t)**. The disadvantage to this notation is that the operator **D** does not work on expressions. This will be important when you verify solutions by substituting expressions for **y(t)**.

The Maple command to solve differential equations is **dsolve**.

Use dsolve *to solve* $dy/dt = t + y$.

```
deq1sol:= dsolve(deq1,y(t));
```

Remember to press Enter (or Return on the Sun) after each statement that ends in a semicolon. The **y(t)** in the **dsolve** command indicates the differential equation is to be solved for $y(t)$. Maple returns the solution with an arbitrary constant, denoted **_C1**.

Whenever it finds an explicit solution, Maple returns an equation with $y(t)$ as the left-hand side and an expression on the right-hand side which gives the solution. It is this expression which is useful for finding values of the solution or generating plots. It is accessed using the **rhs** Maple command.

Use rhs *to access the expression defining the solution.*

```
rh1:=rhs(deq1sol);
```

Notice that only the expression defining the solution is displayed. You can substitute a value for **_C1** into the right-hand side of the solution using the **subs** command and plot the resulting solution for several values of **_C1**.

You can plot several solutions for specific values of the constant.

```
plot({subs(_C1=1,rh1),subs(_C1=2,rh1),
     subs(_C1=0,rh1),subs(_C1=-1,rh1),
     subs(_C1=-2,rh1)},t=-5..5,-5..5);
```

Remember to press Return (or Enter on the Sun) after the first two statements. The graphs of these solutions over the range $-5 < t < 5$ and $-5 < y < 5$ are shown below.

Several solutions to $dy/dt = t + y$.

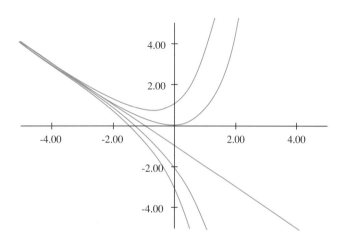

Click on the Go Away box of the graph.

Maple will also solve a wide class of initial value problems. These are differential equations along with an initial value such as $dy/dt = f(t, y)$, $y(t_0) = y_0$.

Solve the initial value problem,
$$\frac{dy}{dt} = \frac{1 - y - \sin(t)}{\cos(t)},$$
$$y(0) = 0.$$

```
deq2:=diff(y(t),t)=(1-y(t)-sin(t))/cos(t);
deq2init:=y(0)=0;
deq2sol:=dsolve({deq2,deq2init},y(t));
```
Note that the curly braces '{' and '}' are used to denote a set of objects in Maple, in this case a set of equations. As a matter of consistency you will always name the equations and the solutions. This is a good habit to develop.

You can plot this solution.

```
plot(rhs(deq2sol),t=-4..4,-3..3);
```
Note that the solution is not valid for all t and that, strictly speaking, the solution for which $y(0) = 0$ is only defined for $-\pi/2 < t < 3\pi/2$. In some cases the command **dsolve** will return an implicit solution, i.e. a relationship between $y(t)$ and t. For example, when solving the equation $dy/dt = -t/y$ Maple gives an implicit solution.

Solve the equation
$dy/dt = -t/y$.

```
deq3:=diff(y(t),t) = -t/y(t);
deq3sol:=dsolve(deq3, y(t));
```
This produces the following output:

$$\mathtt{deq3sol \ := \ y(t)}^2 \ = \ - \ t^2 \ + \ \_C1$$

In some cases where Maple returns an implicit solution, Maple can find an explicit solution using an optional command called **explicit** in the **dsolve** command.

Use dsolve *with the* explicit *option to force Maple to find an explicit solution.*

```
deq3sol2:=dsolve(deq3,y(t),explicit);
```
This produces the following output:

$$\mathtt{deq3sol2 \ := \ y(t) \ = \ - \ (- \ t^2 \ + \ \_C1)^{1/2},}$$
$$\mathtt{y(t) \ = \ (- \ t^2 \ + \ \_C1)^{1/2}}$$

Two solutions are returned in this case which can be accessed as **deq3sol2[1]** and **deq3sol2[2]**. In situations where Maple returns more than one solution for an initial value problem, one has to be careful to check if both solutions work.

You can solve an initial value   `deq3sol3:=dsolve({deq3,y(0)=3},y(t));`
problem when the general
solution is implicit.   If you attempt to verify these solutions, it is obvious that one of them does not satisfy the initial condition. One of the most important topics in differential equations is investigating existence and uniqueness of solutions. In general, if Maple does not find a solution, this does not mean there is no solution. If Maple returns a solution, you still need to verify that it is correct. In the next section you will take a look at differential equations from a graphical point of view to shed some light on what to expect from the solutions.

Using `fsolve` in an Application

A physical example   You will end this section by considering a physical example.
motivates the use of Maple's   Suppose a ball is thrown into the air with an initial velocity
`subs` *and* `fsolve` *com-*   of 9.8 meters/second. How high does it go, and when does
mands.   it hit the ground?

The motion of the ball is governed by Newton's second law of motion, which in this case gives the well-known equation $dv/dt = -g$, where g is the gravitational constant $g = 9.8$ meters/second^2 and v is the velocity. The position of the ball is measured by its height in meters above the ground. Integrating twice, using the initial value for the velocity of 9.8 meters/second, and taking the initial position y to be zero, the following equations for the velocity $v(t)$ and the position $y(t)$ are obtained.

$$v(t) = -9.8t + 9.8$$
$$y(t) = -4.9t^2 + 9.8t$$

To find how high the ball goes, solve for t in the equation $v(t) = 0$ and substitute this value into the equation for $y(t)$. You can verify that $v(1) = 0$ and $y(1) = 4.9$. The ball will hit the ground when $y(t) = 0$ which corresponds to $t = 2$. Using this value for t gives the velocity when the ball hits the ground, $v(2) = -9.8$ meters/second.

Taking air resistance into account provides a better model.

A more realistic model would be to include the force of air resistance. However, you cannot solve the resulting solution for the appropriate values by hand or calculator calculations. Assuming that the air resistance is proportional to the velocity, Newton's second law of motion leads to the equation

$$\frac{dv}{dt} = -g - \frac{r}{m}v.$$

In the equation, m is the mass taken to be 5 grams, g is the gravitational constant 9.8 meters/second^2, r is the coefficient of resistance taken to be 2 grams/second, y is the position, and v is the velocity. You can use Maple to solve this equation for the velocity.

Input and solve the differential equation.

```
bveq:=diff(v(t),t) = -g - r/m*v(t);
bvinit:=v(0)=v0;
bvsol:=dsolve({bveq,bvinit},v(t));
```
The solution is returned explicitly.

Velocity of ball.

```
    bvsol  :=

                      m g                 r t   /m g        \
           v(t)  = - ---  + exp(- ---)    ---  + v0
                      r                 m   \ r        /
```

The position $y(t)$ can be found by solving the differential equation $dy/dt = v(t)$. The Maple expression for $v(t)$ is **rhs(bvsol)**.

Use dsolve *to find the position function.*

```
bpeq:=diff(y(t),t) = rhs(bvsol);
bpinit:=y(0) = y0;
bpsol:=dsolve({bpeq,bpinit},y(t));
```
Again, an explicit solution is returned. A qualitative view of what air resistance does to the flight of the ball is given by the graph of position versus time for the equation with

and without air resistance. In order to plot the solution, you need to substitute values for the constants $g, m, r, v0$, and $y0$.

Plot the position function with and without air resistance.

```
bcons:= g=9.8, m=5, r=2, v0=9.8, y0=0;
plot({subs(bcons,rhs(bpsol)),
      -4.9*t^2+9.8*t},t=0..2,0..5.5);
```

Position function for ball with and without air resistance.

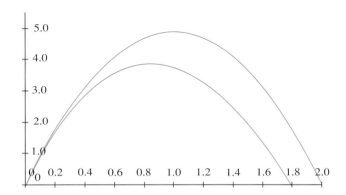

To find the time when the velocity is zero, use Maple's **fsolve** command. You find the maximum height by plugging the result of **fsolve** into the expression for the position.

Time and height of ball when velocity is zero.

```
t1:=fsolve(subs({bcons,v(t)=0},bvsol),t);
maxh:=evalf(subs({t=t1,bcons},bpsol));
```
The height when the velocity is zero is displayed. Similarly, to find the velocity when the ball hits the ground you use Maple's **fsolve** command to find the time when the position is zero. You then substitute this value into the expression for the velocity.

Time and velocity when position is zero.

```
t2:=fsolve(subs({bcons,y(t)=0},bpsol),t);
evalf(subs({bcons,t=t2},bvsol));
```
fsolve gives you a value very close to zero since the position of the ball is zero when $t = 0$. This is not the de-

sired time. You must provide **fsolve** with an approximate range in which to look for the solution. From the graph of the position function, you can see that the position is 0 in the range $1.5 < t < 2$.

Let's try it again with this time range.

```
t2:=fsolve(subs({bcons,y(t)=0},bpsol),
              t,1.5..2);
evalf(subs({bcons,t=t2},bvsol));
```
The velocity at this other time is displayed.

Verifying Solutions

You can use Maple to verify your solutions.

At this point in your course in differential equations, you will have heard the phrase, "You **must** verify your solution by plugging it back into the differential equation", a few dozen times. Using **diff** you can differentiate expressions and using **subs** you can have Maple verify your solutions.

A solution which is difficult to verify by paper-and-pencil methods.

Verify that $y(t) = -e^t + (e^{2t} + 2Ce^{-t})^{1/2}$ is a solution to the differential equation,

$$\frac{y^2}{2} + 2ye^t + (y + e^t)\frac{dy}{dt} = 0.$$

Input the information to verify the solution.

```
eq:=y(t)^2/2+2*y(t)*exp(t)+
          (y(t)+exp(t))*diff(y(t),t)=0;
sol:=-exp(t)+(exp(2*t)+
          2*C*exp(-t))^(1/2);
subs(y(t)=sol,eq);
simplify(");
```
Maple displays $0 = 0$ indicating that the expression substituted for $y(t)$ in the differential equation is indeed a solution. Notice the use of %1 in the substituted equation and its subsequent definition immediately below.

You can verify initial conditions in the same way.

Earlier in this section, trouble occurred when Maple returned more than one solution to an initial value problem. In these situations you can have Maple substitute the initial value into each solution it returns to determine which one you want.

Solve the previous equation with an initial condition and check that the answer has the correct initial condition.

```
Msol:=dsolve({eq,y(0)=2},y(t));
subs(t=0,Msol[1]);
evalf(");
subs(t=0,Msol[2]);
evalf(");
```

Notice that $y(0) = 2$ for one of the solutions and that $y(0) = -4$ for the other. As a double check on Maple itself, you can have Maple verify the solutions it returns.

Plug the solutions back into the original equation.

```
subs(Msol[1],eq);
simplify(");
subs(Msol[2],eq);
simplify(");
```

Again Maple's return of $0 = 0$ indicates these are solutions of the differential equation.

Double checking the computer is just as important or more important than double checking your own work.

Verifying the accuracy of computer algebra systems presents new challenges to the modern-day mathematician. Faced with a problem which can be modeled with a differential equation and a computer generated solution of some sort requires ingenuity to verify the solution. In the coming chapters you will develop many tools to analyze differential equations. The fact that computer algebra systems are being used means the user must develop an even deeper understanding of what is going on to use this new abundance of information.

Additional Activities

Use **dsolve** to find the general solution for each of the initial value problems 1–5. Plot the solution for several values of **C**.

1. $dy/dt = y$

2. $dy/dt = \frac{1}{2}y + t$

3. $dy/dt = t - y$

4. $dy/dt = 5y - 6e^{-t}$

5. $dy/dt = y^2$

6. A ball is thrown into the air with an initial velocity of 10 meters/second. Assuming air resistance proportional to velocity squared with constant of proportionality 0.2, find the maximum height of the ball. As long as the velocity is positive the equation will be

$$\frac{dv}{dt} = -g - \frac{r}{m}v^2$$

assuming that upward is the positive direction.

7. Continuing problem 6 for the velocity negative, the equation will be

$$\frac{dv}{dt} = -g + \frac{r}{m}v^2.$$

Using your results from problem 6 for initial conditions in this equation, find the time and the velocity when the ball hits the ground.

8. Plot the expression for velocity in problem 7 for the range $t = 0..20$. The velocity appears to reach a constant value. Find this value.

5.2 Graphical Procedures

Direction Fields

You can use Maple to get graphical information about a differential equation.

From a geometric point of view, the equation

$$\frac{dy}{dt} = f(t, y)$$

defines the slope of the tangent line at every point in the $t - y$ plane of the solution through that point. You can use Maple to calculate the endpoints of line segments of some fixed length through each point of a grid of points in the

$t-y$ plane, and then you can graph these line segments. The resulting graph is called the direction field of the differential equation.

The Maple procedure to plot a direction field is directionfield.

The Maple procedure used to calculate and plot these line segments is called **directionfield**. This procedure is contained in the **ODE** file which must be read into the current Maple session before it will work. It requires three arguments and has several optional arguments as follows:

General form of direc-tionfield *procedure.*

directionfield(f,h,v,<options>);

f must be a Maple procedure which defines the right-hand side of the equation, **h** is the horizontal range, and **v** is the vertical range of the grid points for which the line segments are calculated.

Plot the direction field for the equation
$dy/dt = y^2 + t^2 - 1.$

```
eq:=(t,y) -> y^2+t^2-1;
directionfield(eq,-2..2,-2..2);
```

Direction field for
$dy/dt = y^2 + t^2 - 1.$

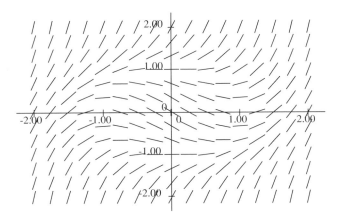

Notice on the graph that the slope of the line segments are negative where $t^2 + y^2 < 1$ and are positive where $t^2 + y^2 > 1$.

Maple will not find a closed form solution for the equation $dy/dt = y^2 + t^2 + 1$. You can, however, use **directionfield** to sketch some approximate solutions. One of the optional arguments for **directionfield** is to specify a set of initial points. A numerical approximation to the solution through each point is drawn using a fourth order Runge/Kutta numerical scheme. Numerical methods such as this are introduced in a later section. In the context of drawing direction fields, these approximate solutions are referred to as flow lines.

You can specify initial conditions for `direction-field`.

```
inits:=[0,0],[0,1],[0,-1],[0,2],[0,-2]:
directionfield(eq,-2..2,-2..2,{inits});
```
Notice the colon after the first statement. As with semicolons, you should press Enter after each statement ending in a colon. Now you can see the flow lines that pass through the points given by the initial conditions.

More `directionfield` *options.*

The option for **directionfield** that controls how many grid points are chosen for drawing the direction field is **grid**. An argument such as **grid=[15,15]** is added to the list of arguments for **directionfield**. To draw a plot that includes flow lines and no line segments of the direction field, just specify **grid=[0,0]**.

In general, the smaller the interval used the more accurate the approximate solution is. The option for **directionfield** that controls the size of this interval is **stepsize**. An argument such as **stepsize=0.1** is added to the list of arguments for **directionfield**. Another option to decrease the step size is **iterations**. For example, **iterations = 5** decreases the step size by a factor of 5 without plotting more points. The routine is much faster if the step size is decreased using the **iterations** option rather than the **stepsize** option, but then the stored points may be too far apart for a nice plot.

Adding flow lines to the directionfield will sometimes give a better indication of what the solutions will look like. Getting the correct step size requires some experimentation, and on a slower machine, it is wise to experiment with a small number of initial values.

Plot the direction field and some flow lines for $dy/dt = -y * \tan(t)$.

```
eq3:=(t,y) -> -y*tan(t);
init3:={[0,1],[0,2],[0,3],[0,4]}:
directionfield(eq3,-6..6,-4..4,init3,
    iterations=5,stepsize=.1);
```

Flow lines for $dy/dt = -y * \tan(t)$.

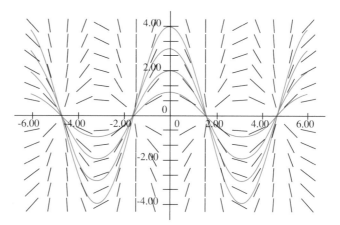

Every solution approaches 0 as t approaches $\pi/2 + n\pi$ for all integers n, as can be verified by using **dsolve** to find the "exact" solution.

Analysis of equations is necessary to use graphical information.

In some cases, **directionfield** allows you to get a rough idea of what solutions to differential equations will look like. There are many parameters and options to set, and in some cases the results will be misleading. Further analysis for some equations is accomplished by considering isoclines. An isocline is a curve in the $t - y$ plane along which the value for dy/dt remains constant or is undefined. Consider the following equation:

$$\frac{dy}{dt} = \frac{y^2 - t^2}{t - 2y}.$$

directionfield *will*
not be useful for drawing
flow lines for many exam-
ples.

If you use **directionfield** with the previous values
for initial conditions the resulting plot will be a mess. This
is due to the fact that for $y = t/2$ the function

$$f(t, y) = \frac{(y^2 - t^2)}{(t - 2y)}$$

is undefined. What happens is that any solution approaching a point on the line $y = t/2$ is discontinuous, and the
Runge/Kutta scheme has no way of detecting this.

The zero isoclines are given by the lines $y = t$ and
$y = -t$. That is, the slope of the direction field is 0 for
any point on these two lines. The zero isoclines and the
undefined isocline break the $t - y$ plane up into 6 regions.
In each region the function $f(t, y) = (y^2 - t^2)/(t - 2y)$ is
either positive or negative throughout the region.

The best way to include the isoclines in a plot would
be to sketch them in by hand after Maple plots the direction field. Another way would be to save the direction
field Maple draws as a named plot structure and then have
Maple sketch the isoclines in a separate plot. You can then
combine the two plot structures using a Maple command
called **display**. This command must be read in with the
command **with(plots)**. You should also put a colon at
the end of each plot statement to avoid having to watch the
entire plot structure scroll across the screen.

Plot the direction field and
some isoclines. Note the
placement of colons.

```
eq2:=(t,y)->(y^2-t^2)/(t-2*y);
eq2field:=directionfield(eq2,-5..5,
                 -5..5,grid=[12,12]):
eq2iso:=plot({-t,t,t/2},t=-5..5,-5..5):
with(plots):
display({eq2field,eq2iso});
```
The graph is shown on the next page.

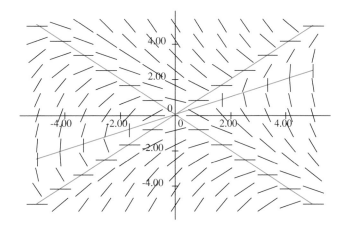

Direction field and some isoclines for
$$\frac{dy}{dt} = \frac{y^2 - t^2}{t - 2y}.$$

You should sketch in some flow lines on the graph by hand.

Additional Activities

Use the **directionfield** procedure to sketch a direction field for each of the following equations. Find the isoclines for each equation and sketch them by hand. Use Maple to sketch flow lines where possible, in other cases sketch them in by hand.

1. $dy/dt = -t/y.$
2. $dy/dt = \sin(4ty).$
3. $dy/dt = 5y - 6e^{-t}.$
4. $dy/dt = t/y.$
5. $dy/dt = y(y - 1)(y + 1).$
6. $dy/dt = t + y.$
7. $dy/dt = t^2 - y^2.$
8. $dy/dt = y + t^2t.$

5.3 Numerical Procedures

Euler's Method

Solutions to differential equations can be approximated.

In "real world" applications of differential equations, solutions are often approximated since closed form solutions cannot always be found. The simplest way to approximate a solution essentially pieces together the field lines you have already drawn for a differential equation. You consider a differential equation of the form

$$\frac{dy}{dt} = f(t, y), \quad y(t_0) = y_0.$$

You can use a numerical algorithm known as Euler's method to approximate solutions to differential equations.

To approximate a solution you calculate the slope of the solution at the initial condition and use this to approximate the solution by a straight line. You then increment t by a fixed amount, calculate the slope of the solution through the resulting point on the line, and then repeat the process. The amount by which t is incremented is called the step size. To a large extent the step size determines how accurate the approximation is. This algorithm is known as Euler's method. The kth iteration of this process is given by the equations

$$t_{k+1} = t_k + h, \quad y_{k+1} = y_k + h f(t_k, y_k)$$

where h is the step size, and $y_0 = y(t_0)$ is the given initial condition.

Maple is a robust programming language specially designed to program mathematics.

If you have had some programming experience, you can see that Euler's method can be easily implemented in many different programming languages. Maple is a simple yet powerful programming language designed to solve mathematics problems. The construction in Maple which is best suited to performing the iterations required by Euler's method is the "for" loop which you have seen in the earlier chapters.

Consider the differential equation $dy/dt = \sin(y)$ with the initial condition $y(0) = 1/3$. You use the variables **tk** and **yk** to store the calculations at each step, and you perform 10 steps with a step size of 0.2.

Use a simple loop in Maple to calculate Euler's method.

```
tk := 0;
yk := evalf(1/3);
for i from 1 to 10
  do
    yk := evalf(yk +.2*sin(yk));
    tk := evalf(tk +.2);
    print(tk,yk);
  od:
```

The colon after the **od** statement (which denotes the end of the **do** statement) suppresses the printing of the calculations as they are executed in the **do** loop. Because Maple does symbolic manipulations, you must use **evalf** to force Maple to do decimal calculations. For this example, some of the **evalf**'s are unnecessary, but it is better to have too many than too few. You may want to try running this example with **evalf** removed from each statement to see for yourself what happens.

It is possible to store the intermediate results in a form in which you can graph them or selectively print out certain values. Maple data structures are flexible and numerous. You will consistently use a data structure known as a **list** to represent a point. For example, the point $(0, 1, -2)$ in three space would be entered as follows:

A Maple list to represent a point.

```
pt := [0,1,-2];
```

The individual components of the point are then accessed via brackets. **pt[1] = 0**, **pt[2] = 1**, and **pt[3] = -2**. You will use a data structure known as an array in Maple to store the points. An array to store the points calculated in the above loop would be declared in the following way:

Declaring an array in Maple.

```
expts := array(0..10);
```

The components of the array are indexed from 0 to 10 and can be accessed via brackets. For example after filling each component of the array **expts** with a point (i.e a **list** of numbers), the first point would be accessed as **expts[0]**, and the second component of the fifth point would be accessed as **expts[4][2]**.

It is a good idea and standard programming practice to initialize some variables before running the loop. This will make it easier to modify the example and run it with different parameters.

Use a loop to program Euler's method in Maple for the equation $dy/dt = y^2 + t^2 - 1$, $y(-2) = -2$ with 20 iterations of step size 0.2.

```
ex := (t,y) -> y^2+t^2-1;
h := 0.2;
n := 20;
tk := -2.0;
yk := -2.0;
expts := array(0..n);
expts[0] := [tk,yk];
for i from 1 to n
  do
    yk:=evalf(yk+h*ex(tk,yk));
    tk:=tk+h;
    expts[i] := [tk,yk];
  od:
print(expts);
```

To plot the points stored in the array **expts**, the array must be converted to a list. This is accomplished with a Maple command called **makelist**.

You can also plot the points stored in the array expts.

```
plot({makelist(expts)});
```

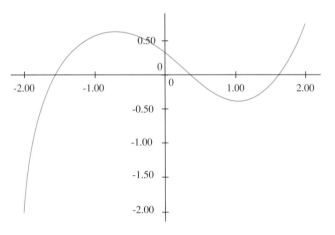

Approximation to solution for $dy/dt = y^2 + t^2 - 1$, $y(-2) = -2$ using Euler's method with step size of 0.2.

Storing the points in an array also provides the advantage of accessing each point using the selection operation for arrays. To print out the last value for y, which in this case represents the approximation to $y(2)$, enter

Print out the approximate value for $y(2)$.

```
expts[20][2];
```

To see how accurate this approximation is, you would reenter the commands with a smaller step size, larger number of iterations, and a *different* name for the array to store the points. You can then subtract the two approximate values to get an idea of how many decimal places of accuracy you have.

Rerun the loop for $dy/dt = y^2 + t^2 - 1$, $y(-2) = -2$ with 40 iterations and a step size of 0.1.

```
ex := (t,y) -> y^2+t^2-1;
h := 0.1;
n := 40;
tk := -2.0;
yk := -2.0;
expts2 := array(0..n);
expts2[0] := [tk,yk];
for i from 1 to n
  do
    yk:=evalf(yk+h*ex(tk,yk));
    tk:=tk+h;
    expts2[i] := [tk,yk];
  od:
print(expts[20][2]-expts2[40][2]);
```

The difference is displayed. Better yet, you can visualize what the change in step size does by plotting both arrays on the same graph along with the direction field. You must use two plot structures, one for the direction field and one for the points you calculated. They are then combined with the **display** command. Remember to use a colon at the end of the plot statements.

Plot the output of the Euler method with different step sizes.

```
with(plots):
exfield:=fieldplot(ex,-2..2,-2..2):
exrkplot:=plot({makelist(expts),
               makelist(expts2)}):
display({exfield,exrkplot});
```

The graph makes it clear that the small difference between the two approximations at $t = 2$ is deceiving.

Additional Activities

Use Euler's method with a step size of $h = 0.1$ to determine an approximate value of the solution at $t = 1$ for each of the initial value problems 1–5. Repeat these computations with $h = 0.05$ and $h = 0.025$ and compare the results with the actual value of $y(1)$. Graph the results along with the direction field in each case.

1. $dy/dt = y$, $y(0) = 1$

2. $dy/dt = y + t$, $y(0) = 1$

3. $dy/dt = t - y$, $y(0) = 1$

4. $dy/dt = 5y - 6e^{-t}$, $y(0) = 1$

5. $dy/dt = 25y(1 - y)$, $y(0) = 1.3$

Use Euler's method with 10, 20, and 40 iterations to determine an approximate value of the solution at the indicated value of t for each of the initial value problems 6–7. Graph the results along with the direction field in each case.

6. $dy/dt = \sin(4ty)$, $y(\frac{\sqrt{\pi}}{2}) = \frac{\sqrt{\pi}}{2}$, $t = \sqrt{\pi}$ (Use the built-in Maple expression **evalf(Pi)** for the value of π.)

7. $dy/dt = y^2$, $y(0) = 1$, $t = 1$

Improved Euler

Euler's method can be improved so as to yield a numerical scheme known as the Improved Euler method. It is similiar to Euler's method in that it is a one step method, but the error is proportional to h^2 where h is the step size. The iterates for $dy/dt = f(t, y)$, $y(t_0) = y_0$ are given by the formula

$$t_{k+1} = t_k + h$$

$$y_{k+1} = y_k + \frac{h}{2}\left(f(t_k, y_k) + f(t_k + h, y_k + hf(t_k, y_k))\right)$$

It is simple to adapt the previous loop to implement this method.

Adapt the previous loop to use the Improved Euler method.

```
ex := (t,y) -> y^2+t^2-1;
h := 0.2;
n := 20;
tk := -2;
yk := -2;
expts3 := array(0..n);
expts3[0] := [tk,yk];
for i from 1 to n
  do
    yk:=evalf(yk+h/2*(ex(tk,yk)+
          ex(tk+h,yk+h*ex(tk,yk))));
    tk:=tk+h;
    expts3[i] := [tk,yk];
  od:
print(expts3);
```

Runge/Kutta

Runge/Kutta is a highly accurate numerical method often used by professional engineers.

Many numerical schemes have been developed. One of the most popular is a fourth order Runge/Kutta method. A Maple procedure to perform the calculations is contained in the **ODE** file.

The error in Runge/Kutta is proportional to h^4, which results in a rapid decrease in errors when the step size is reduced. Due to the accuracy of the method and the fact that each step requires multiple calculations, one must be careful that the round off errors do not become significant, especially for very small step sizes. You will explore this facet of the method at the end of this section.

The ODE *file contains implementations of different numerical schemes for approximating solutions to initial value problems.*

The Euler and Improved Euler methods have also been implemented in Maple procedures. These procedures are called **firsteuler**, **impeuler**, and **rungekutta**. Each of these procedures requires four arguments: the name of a Maple procedure which defines the right hand side of the differential equation, a list of numbers for the initial condition, the step size, and the number of steps to be taken.

Code to use Runge/Kutta to estimate the solution to the differential equation $dy/dt = y, y(0) = 1$ for $0 \le t \le 1$.

```
eq:=(t,y) -> y;
eqrkpts:=rungekutta(eq,[0,1],.2,5);
```
which returns the following:

```
eqrkpts := array(0 .. 5,, [
                0 = [0, 1.]
                1 = [.2, 1.221399999]
                2 = [.4, 1.491817958]
                3 = [.6, 1.822106454]
                4 = [.8, 2.225520824]
                5 = [1.0, 2.718251135]
            ])
```

The exact solution to the problem $dy/dt = y$, $y(0) = 1$ is $y(t) = e^t$. You see that the error is already very small with only 5 iterations.

Calculate the error in Runge/Kutta with 5 iterations.

```
evalf(eqrkpts[5][2]-exp(1));
```

You will end this section with an example to highlight the capabilities of Maple and to explore the possibility of round off errors in the calculations. Return to the problem $dy/dt = y$, $y(0) = 1$. The solution to this equation is $y(t) = e^t$. You estimate the value $y(1)$ using step sizes of 0.1, 0.01, and 0.001 for each method and compare this estimate with the known value of e. With a step size of 0.001 you would be storing 1000 points and you are only interested in looking at the last point. There is an optional fifth argument to each routine which will decrease the step size without storing more points. For example, a 10 in the 5th spot will decrease the step size by a factor of 10 and only every 10th point will be stored in the array. So for each example, you can put a 1 in the third and fourth spot and change the fifth spot appropriately to do more iterations. The calculations are performed as follows:

Compare the output of the numerical methods for the equation
$dy/dt = y, y(0) = 1$
for the value $y(1)$.

```
ex  := (t,y) -> y;
eu1 := firsteuler(ex,[0,1],1,1,10):
eu2 := firsteuler(ex,[0,1],1,1,100):
eu3 := firsteuler(ex,[0,1],1,1,1000):
im1 := impeuler(ex,[0,1],1,1,10):
im2 := impeuler(ex,[0,1],1,1,100):
im3 := impeuler(ex,[0,1],1,1,1000):
rk1 := rungekutta(ex,[0,1],1,1,10):
rk2 := rungekutta(ex,[0,1],1,1,100):
rk3 := rungekutta(ex,[0,1],1,1,1000):
print('euler, impeuler, rungekutta');
print(evalf(exp(1)-eu1[1][2]),
      evalf(exp(1)-im1[1][2]),
      evalf(exp(1)-rk1[1][2]));
print(evalf(exp(1)-eu2[1][2]),
      evalf(exp(1)-im2[1][2]),
      evalf(exp(1)-rk2[1][2]));
print(evalf(exp(1)-eu3[1][2]),
      evalf(exp(1)-im3[1][2]),
      evalf(exp(1)-rk3[1][2]));
```

The output of these print statements is the following:

Errors in Euler, Improved Euler, and Runge/Kutta for step sizes of 0.1, 0.01, and 0.001.

```
euler,          impeuler,        rungekutta
                                           -5
.124539368,  .004200981,   .2081*10
                                           -8
.013467984,  .000044966, -.8*10
                              -6          -7
.001357808,  .357*10,      -.74*10
```

You see that for the Euler method you get one more decimal place of accuracy each time h is divided by 10, and for the Improved Euler method you get two more decimal places of accuracy each time. The Runge/Kutta scheme did not result in 4 more decimal places of accuracy for each division of h by 10. The reason for this is that Maple is doing calculations to 10 decimal places and the Runge/Kutta method should be accurate to 13 digits in the final step of this problem. If you change the value of **Digits** to 20 as follows:

Change the default setting for Digits.

```
Digits:=20;
```

and rerun the computation you get the following results:

Errors in Euler, Improved Euler, and Runge/Kutta for step sizes of 0.1, 0.01, and 0.001 with Digits set to 20.

```
euler,    impeuler,       rungekutta
                                         -5
.124539,  .004200,        .208432*10
                                         -9
.013467,  .000044,        .224643*10
                            -6          -13
.001357,  .452707*10  ,  .226227*10
```

Even with **Digits** set to 10, **rungekutta** is accurate to 5 significant digits with just 10 iterations on this problem. To get this same accuracy with **firsteuler** would require around a million iterations. There is also a scheme called **rungekuttahf** which is identical to **rungekutta** except that it uses floating point calculations tied to the hardware on which Maple is running. The calculations are limited to about 15 digits of accuracy on most machines, which is more than enough for most applications. It is, however, up to 10 times faster and thus for most practical

applications you should use the **rungekuttahf** procedure if you have a floating point processor.

Additional Activities

Use the **rungekuttahf** procedure with a step size of
$h = 0.1$ to determine an approximate value of the solution
at $t = 1$ for each of the initial value problems 1–4. Repeat
these computations with $h = 0.05$ and $h = 0.025$ and compare the results with the actual value. Graph the results
along with the direction field in each case.

1. $dy/dt = y$, $y(0) = 1$
2. $dy/dt = y + t$, $y(0) = 1$
3. $dy/dt = t - y$, $y(0) = 1$
4. $dy/dt = 5y - 6e^{-t}$, $y(0) = 1$
5. A conical tank is 12 meters deep and its open top has a
   radius of 12 meters. Initially the tank is empty. Water is
   added at a rate of 3 meters^3/hour. Water evaporates at a rate
   proportional to the surface area of the water. The constant
   of proportionality is 0.01 meters/hour. Use Runge/Kutta to
   estimate the height of the water to one decimal place accuracy after 100 hours. Plot the depth of the water for the
   first 100 hours. Find the time for which the tank has 6 meters of water in it. To get a starting value for Runge/Kutta,
   assume the evaporation is negligible for the first half hour.

5.4 Picard Iterates

The following theorem is the cornerstone in the study of
differential equations.

Existence and Uniqueness Theorem

*Let f and $\partial f/\partial y$ be continuous in some neighborhood of
(t_0, y_0). Then the initial value problem*

$$\frac{dy}{dt} = f(t, y), \quad y(t_0) = y_0$$

has a unique solution on some open interval containing t_0.

The usual proof of this theorem involves constructing a sequence of functions, known as the Picard iterates of the equation, and showing that they always converge on some interval containing t_0. The Picard iterates are inductively defined by the equations

$$y_1(t) = y_0 + \int_{t_0}^{t} f(s, y_0)ds$$

$$y_2(t) = y_0 + \int_{t_0}^{t} f(s, y_1(s))ds$$

$$\vdots \qquad \vdots$$

$$y_{n+1}(t) = y_0 + \int_{t_0}^{t} f(s, y_n(s))ds.$$

You can use Maple to see how these iterates converge. As you have seen from the Calculus chapter, Maple can compute definite integrals with an indeterminant upper bound, as is needed to compute the Picard iterates. Consider the initial value problem $y' = \sin(t) + y^2, \quad y(0) = 1$.

Compute the first three Picard iterates of $y' = \sin(t) + y^2$, $y(0) = 1$.

```
y0:=1;
pic1:=y0+int(sin(s)+y0^2,s=0..t);
pic2:=y0+int(sin(s)+
    subs(t=s,pic1)^2, s=0..t);
pic3:=y0+int(sin(s)+
    subs(t=s,pic2)^2,s=0..t);
```

Notice that each Picard iterate is an expression in t. You need to substitute this same expression for y in the differential equation with s as the independent variable.

There is a Maple procedure called **picard** built into the **ODE** file which makes computing and plotting several Picard iterates very easy. It requires three arguments — the procedure name which describes the differential equation, an initial point, and the number of iterates to be computed.

Compute and plot the first five Picard iterates and the solution for

$$y' = -\frac{3y}{t+1}, \; y(0) = 1.$$

```
diffeq:=(t,y) -> -3*y/(t+1);
diffeqsol:=dsolve({diff(y(t),t)=
         diffeq(t,y(t)),y(0)=1},y(t));
iterates:=picard(diffeq,[0,1],5);
plot({rhs(diffeqsol),iterates},
                   t=0..3,-2..2);
```

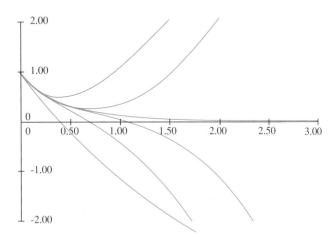

The solution and the first five Picard iterates.

By looking at the form of the solution, it is clear which curve corresponds to the solution because the solution approaches zero as t goes to infinity. Each succeeding iterate is close to the solution for a longer time.

Additional Activities

1. Use the Maple procedure **picard** to compute the first five Picard iterates for the initial value problem

    $$y' = y, \; y(0) = 1.$$

    By hand find the general formula for the nth Picard iterate and verify that the Picard iterates converge to the solution.

2. Use the Maple procedure **picard** to compute the first five Picard iterates for the initial value problem

$$y' = y/t, \ y(1) = 1.$$

Plot these expressions along with the solution $y(t) = t$. By hand find the general formula for the nth Picard iterate. Are the Picard iterates converging to the solution? This example shows clearly why the Picard iterates are not generally a practical method for analyzing differential equations.

5.5 Higher Order Equations and Systems

Maple's **dsolve** command will recognize and solve many higher order differential equations. You first need to know how to translate equations into Maple format. To represent the second derivative you use **(D@@2)**, for the third derivative you use **(D@@3)**, and so on. It is also possible to use **diff(y(t),t,t)** and **diff(y(t),t,t,t)** to represent these derivatives. Since **diff** can be used to differentiate expressions, this will be the preferred notation in some cases. The following are a few examples.

Solve the equation
$y'' + 2y' + 3y = 0.$

```
deq1:= (D@@2)(y)(t)+2*D(y)(t)+3*y(t)=0;
deq1sol:= dsolve(deq1,y(t));
```

The solution returned has two constants, _C1 and _C2, as expected. The form of the solution should also be familiar to you if you have studied second order linear constant coefficient homogeneous differential equations.

Solve the equation
$t^2 y'' + ty' + (t^2 - 1)y = 0.$

```
deq2:=t^2*(D@@2)(y)(t)+t*D(y)(t)
               +(t^2-1)*y(t)=0;
deq2sol:=dsolve(deq2,y(t));
```

The solution is given in terms of two special functions which are built into Maple. These functions are defined in terms of series, and you will explore these a little more later.

Solve the equation
$yy'' + (y')^2 + y' + 1 = 0.$

```
deq3:=y(t)*(D@@2)(y)(t)+(D(y)(t))^2
              +D(y)(t)+1=0;
deq3sol:=dsolve(deq3,y(t));
```
When Maple cannot solve a differential equation it will just return the empty string unless it believes the equation was improperly entered. You will be able to use numerical methods to get approximate solutions to many equations which cannot be solved in closed form.

Initial value problems of higher order can also be solved.

Maple's **dsolve** command will also solve initial value problems of higher degrees. The **D** operator is used to specify the initial conditions as follows.

Solve the initial value problem
$y'' + 2y' + 3y = \sin(t),$
$y(0) = 1/4,$
$y'(0) = -1/4.$

```
deq4:= (D@@2)(y)(t)+2*D(y)(t)
                    +3*y(t)=sin(t);
deq4init:= y(0)=1/4, D(y)(0)=-1/4 ;
deq4sol:= dsolve({deq4,deq4init},y(t));
```
If you work this problem by hand using the method of undetermined coefficients for constant coefficient equations you will get a much simpler looking answer. The answers are the same but Maple uses a different algorithm to come up with a solution.

Maple's **dsolve** can also solve systems of differential equations. The equations are specified in Maple as an expression sequence (i.e., they are separated by commas).

Solve the system
$\frac{dx1}{dt} = -x2$

$\frac{dx2}{dt} = x1.$

```
sys:= D(x1)(t)=-x2(t),D(x2)(t)=x1(t);
sol:=dsolve({sys},{x1(t),x2(t)});
```
Maple returns a set of equations as the solution. The solutions in this case can be accessed as **sol[1]** and **sol[2]**, although you cannot say ahead of time which variable will correspond to 1 and which will correspond to 2.

Let's solve a system of equations with initial values.

```
sys2:=D(x)(t)=y(t),D(y)(t)=
                  -16*x(t)+sin(5*t);
init2:=x(0)=0,y(0)=1;
sol2:=dsolve({sys2,init2},{x(t),y(t)});
```
A pair of equations is returned. There are several interesting graphs to consider for solutions to systems of equations.

Plot x versus t and y versus t for the preceding example.

```
plot({rhs(sol2[1]),rhs(sol2[2])},
    t=0..4*Pi);
```

The two curves correspond to graphs of $x(t)$ and $y(t)$. In particular applications, a parametric plot of $x(t)$ versus $y(t)$ will give you further insight into the system.

Plot a parametric plot of x(t) versus y(t) for the preceding example.

```
plot([rhs(sol2[2]),rhs(sol2[1]),
        t=0..2*Pi]);
```

Compare this graph with the previous graph. The set of points in 3-space of the form $(t, x(t), y(t))$ for t ranging over some interval is referred to as a solution curve of the system. The three curves you just plotted all give different views of a particular solution curve. Maple has another command to plot still more views of a curve in 3-space called **spacecurve**. It is contained in the **plots** package. The notation is similiar to a two-dimensional parametric plot.

Plot a solution curve in 3-space.

```
with(plots):
spacecurve([t,rhs(sol2[1]),rhs(sol2[2])],
    t=0..4*Pi, axes=FRAME);
```

By using the 3-D options you can rotate this curve into any of the three other views you have already plotted.

Homogeneous Linear Systems with Constant Coefficients

Maple's built-in linear algebra package can be useful in solving linear systems with constant coefficients.

An important special case of systems of differential equations are linear homogeneous systems with constant coefficients. Many physical systems are modeled by linear systems. More importantly, the study of stability of general systems of equations is based upon the stability of constant coefficient, linear, homogeneous systems which have the following form:

$$\frac{dx_1}{dt} = a_{11}x_1 + a_{12}x_2 + \ldots + a_{1n}x_n$$

$$\frac{dx_2}{dt} = a_{21}x_1 + a_{22}x_2 + \ldots + a_{2n}x_n$$

$$\vdots \qquad \vdots$$

$$\frac{dx_n}{dt} = a_{n1}x_1 + a_{n2}x_2 + \ldots + a_{nn}x_n$$

It is advantageous to adopt a more compact notation which lends itself to the type of calculations you will perform on the system. The notation is adopted from the linear algebra section of this book.

$$\mathbf{x}(t) = \begin{bmatrix} x_1(t) \\ x_2(t) \\ \vdots \\ x_n(t) \end{bmatrix}, \quad A = \begin{bmatrix} a_{11} & a_{12} & \ldots & a_{1n} \\ a_{21} & a_{22} & \ldots & a_{2n} \\ \vdots & \vdots & & \vdots \\ a_{n1} & a_{n2} & \ldots & a_{nn} \end{bmatrix}$$

With this notation, the system of differential equations is written:

$$\mathbf{x}' = A\mathbf{x} \tag{1}$$

A solution is now given by a vector-valued function. The set of components of a vector-valued solution will be a solution to the original system. Two theorems from the theory of linear homogeneous systems with constant coefficients are needed.

Existence and Uniqueness Theorem for Linear Homogeneous Systems

There exists one, and only one, solution of the initial value problem

$$\mathbf{x}' = A\mathbf{x}, \quad \mathbf{x}(t_0) = \mathbf{x}^0 = \begin{bmatrix} x_1^0 \\ x_2^0 \\ \vdots \\ x_n^0 \end{bmatrix}$$

and this solution exists for $-\infty < t < \infty$.

Vector Space Theorem for Linear Homogeneous Systems

The solutions to a system of n homogeneous linear differential equations with constant coefficients form a vector space of dimension n.

These are very important results. They allow you to search for solutions to the system and give you a criterion for knowing when that search is complete. You begin the search by looking for solutions of the form $\mathbf{x}(t) = e^{\lambda t}\mathbf{v}$, where $\mathbf{v}$ is a constant vector. Observe that

$$\frac{d}{dt}\left(e^{\lambda t}\mathbf{v}\right) = \lambda e^{\lambda t}\mathbf{v}$$

and

$$A(e^{\lambda t}\mathbf{v}) = e^{\lambda t}A\mathbf{v}.$$

It follows that $\mathbf{x}(t) = e^{\lambda t}\mathbf{v}$ is a solution if and only if

$$e^{\lambda t}A\mathbf{v} = \lambda e^{\lambda t}\mathbf{v}.$$

After dividing by $e^{\lambda t}$ you find that λ and $\mathbf{v}$ must satisfy the equation $A\mathbf{v} = \lambda\mathbf{v}$. Recall that in this case $\mathbf{v}$ is known as the eigenvector associated to the eigenvalue λ for the matrix A. Also, recall that λ is a root of the characteristic polynomial of A. The eigenspace of λ is the subspace of all the eigenvectors associated to λ. You saw in the chapter on linear algebra that Maple has a built-in facility for finding eigenvectors and eigenvalues of matrices. The reader should refer back to the chapter on linear algebra for entering matrices and finding eigenvalues and eigenvectors in Maple. Consider the following example:

$$\mathbf{x}' = \begin{pmatrix} 1 & -1 & 4 \\ 3 & 2 & -1 \\ 2 & 1 & -1 \end{pmatrix} \mathbf{x} \tag{2}$$

Use Maple to find the eigenvectors of the matrix.

```
with(linalg):
A:=matrix(3,3,[1,-1,4,3,2,-1,2,1,-1]);
eigsA:=eigenvects(A);
```

In this case, Maple returns the following seqence of 3 lists.

```
eigsA := [1, 1, {[ -1, 4, 1 ]}],
         [-2, 1, {[ -1, 1, 1 ]}],
         [3, 1, {[ 1, 2, 1 ]}]
```

Recall that the first component of each list is an eigenvalue, the second component is the multiplicity of the eigenvalue for the characteristic polynomial of A, and the third component is a set of eigenvectors which span the eigenspace of the eigenvalue. In this case you found three eigenvalues each with an eigenspace of dimension one. From a basic theorem of an advanced linear algebra course the three eigenvectors are linearly independent. From this it follows that the solutions constructed from the three eigenvalues will be linearly independent. Therefore, by the Vector Space Theorem, you can construct the general solution from these three solutions.

Obtain the general solution to equation (2).

```
solA:=
  C1*exp(eigsA[1][1]*t)*eigsA[1][3][1]
 +C2*exp(eigsA[2][1]*t)*eigsA[2][3][1]
 +C3*exp(eigsA[3][1]*t)*eigsA[3][3][1];
solA:= evalm(");
```

You can compare this solution with the solution generated by Maple's **dsolve** operator.

Find the general solution to equation (2) using dsolve.

```
eq1:=D(x1)(t)=x1(t)-x2(t)+4*x3(t);
eq2:=D(x2)(t)=3*x1(t)+2*x2(t)-x3(t);
eq3:=D(x3)(t)=2*x1(t)+x2(t)-x3(t);
dsolve({eq1,eq2,eq3},{x1(t),x2(t),x3(t)});
```

The answers look different, but you should be able to verify that they do indeed give the same general solution.

Complex eigenvalues complicate matters somewhat.

If λ is a complex eigenvalue with complex eigenvector **v** then $\mathbf{x}(t) = e^{\lambda t}\mathbf{v}$ will be a complex solution to the differential equation. It follows from the fact that complex solutions come in conjugate pairs that the real and imaginary parts of a complex solution are real solutions. Maple has built-in facilities **Re** and **Im** for extracting real and imaginary parts of complex numbers.

Find the eigenvectors of the matrix.

```
B:=matrix(3,3,[1,2,-1,0,1,1,0,-1,1]);
eigsB:=eigenvects(B);
```
When you enter the previous statements, you get the following output. Note that the order of the eigenvalues may be different each time you enter the equations.

```
   eigsB := [1, 1, {[ 1, 0, 0 ]}],
                        2
    [RootOf(2 - 2 _Z + _Z ), 1,
                                  2
      {[ - 2 RootOf(2 - 2 _Z + _Z ) + 1, 1,
                                  2
           - 1 + RootOf(2 - 2 _Z + _Z ) ]}
    ]
```

The complex eigenvalues are given in terms of Maple's **RootOf** function. As you saw in the linear algebra chapter, Maple can sometimes convert these to radical form with the use of the Maple command **allvalues**. The order of the terms of the output in the previous statements is important in what follows.

Construct a solution from the real eigenvalue.

```
sol1:=exp(eigsB[1][1]*t)*eigsB[1][3][1];
```
Follow what was done in the linear algebra chapter for the complex eigenvalues.

Substitute one component of the result of allvalues *for the* RootOf *expression.*

```
complexeig:=allvalues(eigsB[2][1]);
comvect:=subs(eigsB[2][1]=complexeig[1],
    eigsB[2][3]);
comsol:=exp(complexeig[1]*t)*comvect[1];
```

You can now use **Re** and **Im** to find two real solutions from this complex solution, and use them to construct the general solution.

Use Re and Im to obtain
two real vector value
solutions.

```
sol2:=Re(comsol);
sol3:=Im(comsol);
```

The solutions which are returned do not look like real-valued functions because Maple does not automatically distribute multiplication over the vectors and does not simplify complex numbers automatically.

Construct the general
solution and use evalm
and evalc *to evaluate the*
answer.

```
solB:=evalm(C1*sol1+C2*sol2+C3*sol3);
solB:=map(evalc,solB);
```

The general vector valued solution to the differential equation is returned.

You still get n solutions of the form $\mathbf{x}(t) = e^{\lambda t}\mathbf{v}$ if there are less than n eigenvalues but there are n linearly independent eigenvectors. This happens when the dimension of the λ-eigenspace is equal to the multiplicity of λ as a root of the characteristic polynomial for each eigenvalue.

If the dimension of one or more of the eigenspaces is less than the multiplicity of the eigenvalue as a root of the characteristic polynomial, there will not exist n linearly independent solutions of the form you are looking for. A general approach to this situation is to look for solutions of the form $\mathbf{x}(t) = e^{At}\mathbf{v}$ where $\mathbf{v}$ is a constant vector and $e^{At}\mathbf{v}$ is defined as follows:

$$e^{At} = I + At + A^2\frac{t^2}{2!} + ... + A^n\frac{t^n}{n!} + ...$$

The above infinite series always converges, although in general you cannot calculate it. Since the series converges, it can be differentiated term by term and after some simplification you get

$$\frac{d}{dt}(e^{At}\mathbf{v}) = Ae^{At}\mathbf{v}.$$

It then follows that $e^{At}\mathbf{v}$ is a solution of equation (1) for every constant vector $\mathbf{v}$. The general solution will be sums of terms of the form

$$e^{\lambda t}\left(\mathbf{v}^1 + t\mathbf{v}^2 + t^2\mathbf{v}^3 + ... + t^{n-1}\mathbf{v}^n\right)$$

where n is the multiplicity of λ as a root of the characteristic polynomial and $\mathbf{v}^1, .., \mathbf{v}^n$ are constant vectors.

Computing e^{At} when the eigenvalues of A can be found.

It turns out that n linearly independent vectors $\mathbf{v}$, can be found for which the infinite series $e^{At}\mathbf{v}$ can be summed exactly if the eigenvalues of A can be computed. You will not explore this fact here, and, in fact, it is beyond the scope of most introductory differential equations texts. It depends on the Cayley-Hamilton theorem which was introduced in the chapter on linear algebra.

You must use **allvalues** to see if Maple can find the eigenvalues exactly in the case where Maple returns the eigenvalues in the **RootOf** form. It will return decimal approximations in cases where it cannot. Making use of approximate solutions obtained from approximated eigenvalues is the subject matter of an entire course which will not be covered here.

Maple has a built-in facility for finding e^{At}

It turns out that if the eigenvalues of a matrix can be found exactly then e^{At} can be calculated. You can then use e^{At} to construct the general solution. For many matrices, A, Maple can perform this computation. **dsolve** will not work for some of these examples (at least not in a reasonable amount of time) because it uses more general algorithms which work on nonlinear equations and may not be efficient for homogeneous linear systems. The Maple command is **exponential** and it takes two arguments, a matrix and a variable.

Consider an example where eigenspaces are not sufficient to find all solutions.

```
C:=matrix(3,3,[1,1,0,0,1,0,0,0,2]);
expCt:=exponential(C,t);
solC:=C1*expCt &* [1,0,0] +
        C2*expCt &* [0,1,0] +
        C3*expCt &* [0,0,1];
solC:=evalm(solC);
```

Notice that the answer contains a factor of t as one expects when there are not enough eigenvectors to form a solution which only contains exponentials.

Additional Activities

In exercises 1–5 find the eigenvalues and eigenvectors for the matrix, A, and use these to find the general solution to $\mathbf{x}'(t) = A\mathbf{x}(t)$ where possible. In other cases use e^{At} to construct the general solution. In each case, where possible, compare with the solution returned by **dsolve**.

1. $A = \begin{bmatrix} 1 & 0 & 0 \\ 0 & 2 & 0 \\ 0 & 0 & 3 \end{bmatrix}$

2. $A = \begin{bmatrix} 3 & 2 & 4 \\ 2 & 0 & 2 \\ 4 & 2 & 3 \end{bmatrix}$

3. $A = \begin{bmatrix} 1 & -2 & 1 \\ 0 & -2 & 1 \\ -1 & 3 & 0 \end{bmatrix}$

4. $A = \begin{bmatrix} 1 & 1 & 1 \\ 2 & 1 & -1 \\ -3 & 2 & 4 \end{bmatrix}$

5. $A = \begin{bmatrix} 2 & 0 & -1 & 0 \\ 0 & 2 & 1 & 0 \\ 0 & 0 & 2 & 0 \\ 0 & 0 & -1 & 2 \end{bmatrix}$

6. Convert the system

$$x'' + 2x - y = 0$$
$$y'' + 2y - x = 0$$

into a linear system of 4 equations by making the substitutions $x_1(t) = x(t), x_2(t) = x'(t), x_3(t) = y(t), x_4(t) = y'(t)$. Find the eigenvalues and eigenvectors of the resulting system and use them to find the general solution to the system. Find a particular solution for which $x(0) = 0$, $x'(0) = 16$, $y(0) = 0$, $y'(0) = 0$. Make a parametric plot of $x(t)$ vs $y(t)$ for this particular solution.

5.6 Phase Space

The eigenvalues associated with a system of linear homogeneous differential equations give clues to the qualitative behavior of the solutions to the differential equation. It is clear, for instance, that if there are n distinct eigenvalues for a system of n equations and they are all negative, that all solutions tend to 0. The goal in this section is to get some qualitative (graphical) information for systems of two first order equations.

A general system of two first order equations can be written as follows.

$$\frac{dx_1}{dt} = f_1(t, x_1, x_2)$$
$$\frac{dx_2}{dt} = f_2(t, x_1, x_2). \tag{1}$$

Solutions are curves in three-dimensional space.
As you saw before, solutions to Equation (1) parametrically describe a curve in three space. This solution curve is the set of points $(t, x_1(t), x_2(t))$ as t ranges over some interval for which the solution is defined. Consider the system of equations

$$x_1' = x_2, \quad x_2' = -x_1 - \frac{1}{2}x_2, \quad x_1(0) = 0, \quad x_2(0) = 1.$$

Solve the system of equations and plot a solution curve.

```
deq1:=D(x1)(t)=x2(t);
deq2:=D(x2)(t)=-x1(t)-x2(t)/2;
deqinit:=x1(0)=0,x2(0)=1;
deqsol:=dsolve({deq1,deq2,deqinit},
         {x1(t),x2(t)});
with(plots):
spacecurve(
   [t,rhs(deqsol[1]),rhs(deqsol[2])],
          t=0..20,   axes=NORMAL);
```

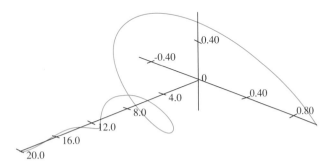

Solution curve plotted with spacecurve *command.*

The system of equations in this example has two negative eigenvalues, and you can see from the plot that this solution is approaching the t-axis as t goes to infinity. That is, both x_1 and x_2 are approaching 0 for this curve.

Direction fields are not practical for three-dimensional systems.

The analog of the direction field you drew for one equation would be three-dimensional and would be rather difficult to visualize on a two-dimensional computer screen. As it turns out, many physically interesting differential equations are independent of time (i.e., f_1 and f_2 in equation (1) do not depend on the t variable). These are known as **autonomous differential equations**.

For the autonomous case of two first order equations, the functions f_1 and f_2 determine a vector for each point in the $x_1 - x_2$ plane. The resulting vectors are known as phase vectors, and they completely determine the behavior of the solutions to an autonomous differential equation. A curve in the $x_1 - x_2$ plane traced out by a solution is called an orbit of the solution. If f_1, f_2, and their partial derivatives are continuous in a region around a point in the $x_1 - x_2$ plane, then there will be a unique orbit through that point. The $x_1 - x_2$ plane is known as the phase plane or phase space of the differential equation. A picture of the orbits and phase vectors is known as a phase portrait.

The `phaseplot` *procedure calculates and plots phase vectors and orbits.*

To get a picture of the phase vectors, a grid of points in the $x_1 - x_2$ plane is chosen, and for a point (a, b) on the grid, the vector, $(f_1(0, a, b), f_2(0, a, b))$, is plotted with its tail at the point (a, b). Maple produces a plot of these phase vectors with a call to the **phaseplot** procedure. Note each vector is scaled by the same amount so that the vectors do not run into each other in the plot.

Plot the phase vectors for the previous differential equation.

```
eq1:=(t,x1,x2)->x2;
eq2:=(t,x1,x2)->-x1 - x2/2;
phaseplot([eq1,eq2],-2..2,-2..2);
```

Phase portrait of
$$x_1' = x_2$$
$$x_2' = -x1 - x_2/2.$$

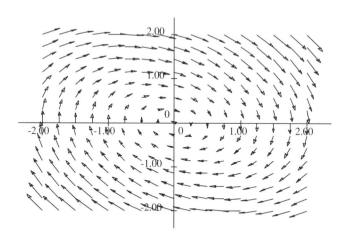

The phase portrait for this system is similiar to the direction field of the first order differential equation

$$\frac{dx_2}{dx_1} = \frac{-x_1 - x_2/2}{x_2}.$$

The slopes of the phase vectors will be equal to the slopes of the line segments of the direction field.

You can combine this graph with a plot of the orbit of the solution you found earlier. You must be careful to plot x_1 along the horizontal axis and x_2 along the vertical axis if that is the order the variables were specified for the **phaseplot** command.

Plot the solution you found earlier along with the phase vectors.

```
with(plots):
orb1:=plot([rhs(deqsol[2]),rhs(deqsol[1]),
    t=0..20],-1..1,-1..1):
phasevects:=phaseplot([eq1,eq2],
    -1..1,-1..1):
display({orb1,phasevects});
```

Compare this with the **spacecurve** you plotted earlier. By rotating the earlier solution curve appropriately you can make it match the picture of the orbit in the phase portrait.

phaseplot *can be used to plot orbits.*

For many examples, a "closed form" solution cannot be found. **phaseplot** has a facility for plotting orbits similiar to the way in which **fieldplot** was able to approximate solutions using the Runge/Kutta numerical method. The options to **phaseplot** are similiar to **fieldplot** in that initial points can be specified for orbits and the options **grid**, **iterations**, and **stepsize** are implemented in the same way. There is an additional argument called **numsteps** to specify how many steps are taken forward and backward by the Runge/Kutta numerical scheme which calculates the orbits.

Consider the following second order differential equation which comes from Newton's second law of motion for a pendulum with friction:

$$\frac{dy^2}{d^2t} = -\sin(y) - 0.2\frac{dy}{dt}$$

where y is the radian measure of the pendulum from the bottom position. This leads to a system of two first order equations:

$$dy/dt = v$$
$$dv/dt = -\sin(y) - 0.2v.$$

Plot the phase portrait and some orbits for the "pendulum with friction" equations.

```
pend1:=(t,y,v) -> v;
pend2:=(t,y,v) -> -sin(y) - 0.2*v;
pendinits:=[0,2],[0,2.5],[0,3],[0,3.5]:
phaseplot([pend1,pend2],-10..10,-6..6,
          {pendinits},numsteps=200);
```

Phase portrait and orbits for pendulum equations.

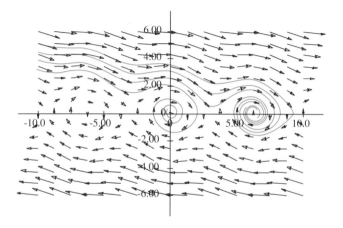

The points $(n\pi, 0)$ for any integer n are points where the phase vectors have length 0, and are called equilibrium points. From this you can see that any solution which comes near the points $(2n\pi, 0)$ tend to approach these points. The points $((2n + 1)\pi, 0)$ are unstable equilibrium points. The study of equilibrium points is an important topic in theory of autonomous systems. Phase portraits are a good tool to help understand these and other important characteristics of autonomous systems.

Additional Activities

Transform each of equations 1–5 into a system of two first order equations by making the sustitution $v = dy/dt$. Use Maple's **phaseplot** procedure to draw a phase portrait for each equation. In each case, find the regions in which dy/dt and dv/dt are positive and negative, and compare this information with the phase portraits drawn by Maple. How do the eigenvalues of the associated system of equations affect the orbits?

1. $y'' + y' - 2y = 0$
2. $y'' - 3y' + 2y = 0$
3. $y'' - 2y' + y = 0$
4. $y'' - 2y' + 2y = 0$
5. $y'' - 2y' - 2y = 0$

In exercises 6–10 use Maple to draw a phase portrait. For each equilibrium point, decide if the nearby orbits are approaching the equilibrium point.

6. $x' = y^2 - x^2,\ y' = x - 2y$
7. $x' = y^2 - x^2,\ y' = x - \sin(y)$
8. $x' = \sin(y) - x,\ y' = \cos(x) - y$
9. $x' = y^2 + x^2 - 4,\ y' = y^2 - x^2$
10. $x' = \sin(2*x) - y,\ y' = \cos(2*x) - y$

5.7 Numerical Methods for Systems

Runge/Kutta

Numerical methods for a system of first order differential equations are no harder to implement than those for a single first order equation. However, the points are generated for higher dimensional coordinate systems. A fourth order Runge/Kutta scheme is used to get the orbits in the

phaseplot procedure. You can use the **rungekutta** and **rungekuttahf** procedures to estimate solutions to a system of up to fifty first order equations. You enter the equations in the same way only they are grouped in a list, just as the initial conditions are. A call to this procedure for a system of two first order equations might look like the following:

```
f:=(x,y,z) -> <expression in x,y,z>
g:=(x,y,z) -> <expression in x,y,z>
rungekuttahf([f,g],[x0,y0,z0],h,n):
```

This procedure would return an array indexed from 0 to n. Each array entry contains a list of three numbers which represents a point in three-space. You can plot these points two coordinates at a time using a Maple procedure called **makelist** and is written **makelist(A,m,n)** where **A** is an array of lists of numbers and **m** and **n** are integers which denote the position of the coordinates you wish to plot. For example, **makelist(A,2,3)** will pick out the second and third coordinates of each point in the array **A**. Consider the following pair of second order equations. They represent a double pendulum, that is a pendulum attached to a pendulum.

$$\theta_1'' = -2\sin(\theta_1) + \sin(\theta_2)$$
$$\theta_2'' = -2\sin(\theta_2) + 2\sin(\theta_1)$$

You can translate this into a system of four first order equations using the variables $x = \theta_1$, $xp = \theta_1'$, $y = \theta_2$, and $yp = \theta_2'$. Assume the pendulums start at rest and the bottom pendulum is given a push of 1/2 unit/second.

Solve the double pendu-
lum equations.

```
dp1:=(t,x,xp,y,yp)-> xp;
dp2:=(t,x,xp,y,yp)-> -2*sin(x)+sin(y);
dp3:=(t,x,xp,y,yp)-> yp;
dp4:=(t,x,xp,y,yp)-> -2*sin(y)+2*sin(x);
dpinit:=[0,0,0,0,1/2];
dppts:=rungekuttahf([dp1,dp2,dp3,dp4],
          dpinit,.1,300):
```

The colon at the end of the **rungekuttahf** statement will suppress the printing of all 300 points. In some cases you will want to look at the array of points generated and can add a print statement to do so. In this case, you look at a plot of the approximate solution for θ_1 vs θ_2.

Plot the second and fourth
coordinates.

```
plot({makelist(dppts,2,4)});
```

A plot of the solution to the
system for θ_1 vs θ_2 gener-
ated using rungekuttahf.

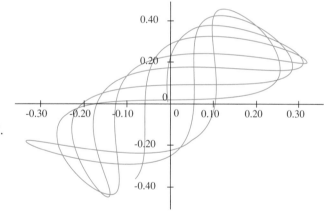

The point of this example is to illustrate the use of **runge-kuttahf** for systems of equations. Deeper analysis of this equation would begin with linearizing the equation and studying the resulting system.

Additional Activities

1. A mass–spring–dashpot system is governed by the second order equation $my'' + cy' + ky = 0$, where m is the mass, c is the damping constant, and k is the spring constant. Use **rungekuttahf** to draw the solutions to the equation in the $y-y'$ plane for $k = 1$, $m = 1$, and $c = 0, 0.5, 1.0, 1.5, 2$. Use $y'(0) = 1, y(0) = 0$ for your initial values and do 30 iterations with a step size of 0.5 for each.

2. A ball of mass 1 gram is thrown into the air with an initial velocity of 10 meters/second. Assuming air resistance proportional to velocity with constant of proportionality 2.0, use **rungekuttahf** to find the maximum height of the ball and the velocity of the ball when it hits the ground. The height of the ball satisfies the second order equation

   $$my'' = -mg - ry'$$

   assuming that upward is the positive direction.

3. Use the same setup as problem 2 only assume air resistance is proportional to velocity squared with constant of proportionality 0.2. For positive velocity the equation is

   $$my'' = -mg - r(y')^2$$

   and for negative velocity the equation is

   $$my'' = -mg + r(y')^2.$$

   Plot position versus time for the cases where air resistance is proportional to velocity and velocity squared and the case where it is ignored.

4. A long heavy string is attached to a balloon filled with helium. The motion of the balloon is governed by Newton's equation which gives

$$\frac{d}{dt}(mv) = mg + H + R$$

where y is the height of the balloon above the ground, m is the mass of the part of the string which is above the ground (and is thus a function of position), H is the force of helium on the balloon, R is the force of air resistance, and g is the gravitational constant. Suppose that ρ is the linear density of the string so that $m = \rho y$, $R = \lambda v$ is the force of air resistance where λ is the resistance coefficient, and y_e is the equilibrium position of the balloon (i.e. $H = y_e \rho g$).

Consider the situation where $g = 9.8$ meters/second^2, $\rho = 0.25$ kilograms/meter, $\lambda = 0.03$ kilograms/second, and $y_e = 2.0$ meters. Use the **phaseplot** procedure in Maple to sketch the phase portrait for this setup. The balloon is pulled down to a height of 30 centimeters and released. Use the **rungekuttahf** procedure in Maple to predict the highest point the balloon will reach and the time at which it will reach this height.

5. If the ball in problem 1 also has a horizontal position denoted by x, and assuming the air resistance is proportional to velocity directed opposite to its instantaneous direction of motion, you get the following equations:

$$mx'' = -rx'\sqrt{(x')^2 + (y')^2}$$

$$my'' = -ry'\sqrt{(x')^2 + (y')^2} - mg.$$

Assume a trajectory of 45 degrees, initial velocity of 10 meters/second, and coefficient of resistance of 2. Use **rungekuttahf** to find the maximum height of the ball, the time when the ball hits the ground, and the horizontal distance traveled. Compare these quantities with the flight of the ball without air resistance. Plot the solution in the $x - y$ plane with and without air resistance.

6. An iron mass is attached to a spring and is suspended above a magnet. The spring exerts a force on the iron mass proportional to the distance the spring is stretched/compressed from its natural length with proportionality constant κ. The magnet exerts a force inversely proportional to the distance between the iron mass and the magnet with proportionality constant μ. Taking $y = 0$ to be the position of the mass when the spring force is equal to the weight of the mass and taking d to be the distance to the magnet from this position, derive the following equation:

$$my'' = -\kappa y + \frac{\mu}{d - y}.$$

Using the values $m = 1$, $\kappa = 1$, $\mu = 20$, and $d = 10$, use **phaseplot** to sketch the phase plane for $-1 < y < 9$ and $-4 < v < 4$. Use **rungekuttahf** to sketch the motion for y vs v if the mass is pulled down to $y = 7$ and released. Find the two points in the phase space where the system is in equilibrium and describe the motion for initial values near these two equilibria.

5.8 dsolve **Options**

Laplace Transforms

There are three options for the Maple **dsolve** operator beside the option **explicit**. These are **laplace**, **series**, and **numeric**. Each of these methods will be effective on different types of differential equations. The laplace option is well suited for equations which have the following form:

$$ay'' + by' + cy = f(t), \ y(0) = y_0, \ y'(0) = y_0'.$$

The function, $f(t)$, is referred to as the forcing function.

Use the laplace *option to solve* $y'' + 2y' - y = \sin(t)$.

```
lapex:=(D@@2)(y)(t)+2*D(y)(t)-
          y(t) = sin(t);
lapexsol:=dsolve(lapex,y(t),laplace);
```
Notice that the answer contains **y(0)** and **D(y)(0)** as constants. If initial conditions for this method are speci-

fied they must be at 0. Solving this equation without the laplace option results in a very different looking answer since Maple uses a different algorithm to solve the equation. It is difficult to verify that the solutions are the same, and Maple is not able to simplify the difference of the two answers to 0. However, you can verify that both answers are actually solutions, and the uniqueness theorem applies in this case to prove they are the same.

Compare the solutions with and without the `laplace` *option.*

```
lapex2:=diff(y(t),t,t)+2*diff(y(t),t) -
        y(t) = sin(t);
init2:= y(0)=1,D(y)(0)=1;
sol1:=dsolve({lapex2,init2},y(t),laplace);
sol2:=dsolve({lapex2,init2},y(t));
simplify(rhs(sol1)-rhs(sol2));
simplify(subs(sol1,lapex2));
simplify(subs(sol2,lapex2));
```

Notice that even though the difference of the two solutions did not simplify to zero, they are both solutions as the last two statements returned $\sin(t) = \sin(t)$. They must, therefore, be equal.

You can compute the Laplace transform and the inverse Laplace transform.

The **laplace** option for **dsolve** is based on computing Laplace transforms. Maple has a built-in procedure for computing Laplace transforms called **laplace**. To use it to solve an equation, the following steps are used:
1. Laplace transform the entire equation
2. Solve for the Laplace transform of the dependent variable
3. Inverse Laplace transform the resulting equation.

Either before step (1) or after step (3) you will want to plug in the initial condition.

Solve
$y'' + y' + 3y = \cos(t)$,
$y(0) = 2, y'(0) = -1$
using Laplace transforms.

```
eq:=diff(y(t),t,t)+diff(y(t),t)
      + 3*y(t)=cos(t);
Leq:=laplace(eq,t,s);
LY:=solve(Leq,laplace(y(t),t,s));
sol:=invlaplace(LY,s,t);
subs({y(0)=2,D(y)(0)=-1},sol);
```

The solution is displayed. You should compare this answer with the answer you get using **dsolve(...,laplace)**.

Heaviside Function

The method of Laplace transforms is especially applicable in cases where the function, $f(t)$, is discontinuous. Many discontinuous functions can be expressed in terms of a simple function known as the Heaviside function, $H(t)$, which is defined as follows.

$$H(t) = \begin{cases} 0, & \text{if } t < 0; \\ 1, & \text{if } t \geq 0. \end{cases}$$

Maple has the function **Heaviside** built into its library. This function must be read into each Maple session in which it is used as follows:

Make Heaviside *available for calculations.*

```
readlib(Heaviside);
```

The laplace option for **dsolve** will recognize and solve equations which involve the Heaviside function. Consider the following example:

$$y'' + 3y' + 2y = \begin{cases} 1, \text{ if } 0 \leq t \leq 1 \\ 0, \text{ if } t > 1 \end{cases}, \; y(0) = 0, \; y'(0) = 1.$$

This can be written in terms of the Heaviside function as follows:

$$y'' + 3y' + 2y = H(t) - H(t-1), \; y(0) = 0, \; y'(0) = 1.$$

Solve this example.

```
hex:=(D@@2)(y)(t)+3*D(y)(t)+2*y(t)=
    Heaviside(t)-Heaviside(t-1);
hinit:=y(0)=0,D(y)(0)=1;
hexsol:=dsolve({hex,hinit},
    y(t),laplace);
```

The answer, as expected, is given in terms of the Heaviside function. The solution is nevertheless a continuous function as you can see by plotting the solution.

Sketch the solution.

```
plot(rhs(hexsol),t=0..10);
```

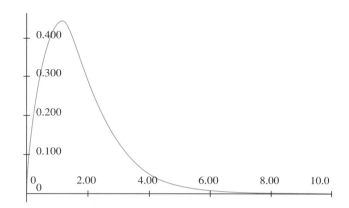

Solution to
$y'' + 3y' + 2y = H(t) - H(t-1),$
$y(0) = 0, \ y'(0) = 1.$

It is common in applications to have more complex discontinuous functions. Consider the function, $g(t)$, given by the graph:

Example of discontinuous function.

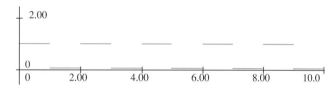

This function can be written using the Heaviside function as follows:

$$g(t) = \sum_{n=0}^{n=4} H(t - 2n) - H(t - 2n - 1).$$

You use the **sum** command to express this function and then solve the equation $y'' + 3y + 2y = g(t)$.

Solve this equation and graph the solution.

```
n:='n';
f:=sum(Heaviside(t-2*n)-
            Heaviside(t-2*n-1),n=0..4);
eq:=(D@@2)(y)(t)+3*D(y)(t)+2*y(t) = f;
eqi:=y(0)=0,D(y)(0)=1;
eqsol:=dsolve({eq,eqi},y(t),laplace);
plot(rhs(eqsol),t=0..10);
```

Solution to
$y'' + 3y' + 2y = g(t),$
$y(0) = 0, \; y'(0) = 1.$

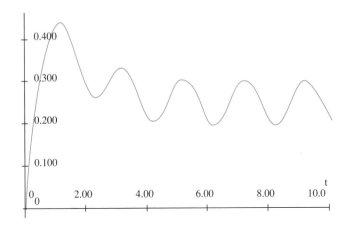

Again you see that the solution appears to be continuous.

Additional Activities

Use **dsolve** with the **laplace** option to solve the following initial-value problems. Plot the solutions for the interval $0 \leq t \leq 10$.

1. $y'' + y = \begin{cases} 0, & 0 \leq t \leq \pi \\ 1, & \pi < t \end{cases}$; $y(0) = 0, \; y'(0) = 1$

2. $y'' + 2y' + y = \begin{cases} t, & 0 \leq t \leq 1 \\ 0, & 1 < t \end{cases}$; $y(0) = -1, y'(0) = 1$

3. $y'' + y' + 7y = \begin{cases} t, & 0 \leq t \leq 2 \\ 0, & 2 < t \end{cases}$; $y(0) = 0, y'(0) = 0$

4. $y'' + y = \sum_{n=0}^{4} \left(H(t - 2n\pi) - H(t - (2n+1)\pi) \right)$
$y(0) = 0, y'(0) = 0$

Maple also recognizes the Dirac delta function as **Dirac**. Type **?Dirac** for more information. The usual mathematical notation is $\delta(t)$.

5. $y'' + y = \sum_{n=0}^{4} \delta(t - 2n\pi), y(0) = 0, y'(0) = 0$

6. $y'' + y = \sum_{n=0}^{8} \delta(t - n\pi), y(0) = 0, y'(0) = 0$

Series Solutions

Next you will consider general linear differential equations of the form:

$$a(t)y''(t) + b(t)y'(t) + c(t)y(t) = f(t)$$

Series solutions offer a way to handle some of these equations. If $a(t), b(t), c(t)$, and $f(t)$ have convergent power series in some interval around a point t_0, then you might expect that the solution has a convergent power series on some interval around t_0.

If $a(t_0) \neq 0$ and $a(t), b(t), c(t)$, and $f(t)$ can be expressed as power series around t_0, then t_0 is called an ordinary point of the differential equation. In this situation, a power series convergent on some interval containing t_0 is given by the Taylor polynomial:

$$y(t) = \sum_{n=0}^{\infty} y^{(n)}(t_0) \frac{(t - t_0)^n}{n!}.$$

To find the terms in the above power series, you calculate the derivatives of y at t_0. Consider the following equation:

$$(1 - t^2)y'' - 6ty' - 4y = 0, \quad y(0) = 0, \quad y'(0) = 1$$

Look for the first few terms of a series solution near the ordinary point $t = 0$. You find $y''(0)$ by solving the equation for $y''(t)$ and then substituting the value $t = 0$ into the equation.

Find $y''(0)$.

```
sereq:=(1-t^2)*diff(y(t),t,t)-
              2*t*diff(y(t),t)+4*y(t)=0;
yppt:=solve(sereq,diff(y(t),t,t));
ypp0:=subs({t=0,y(t)=0,diff(y(t),t)=1},
              yppt);
```

The second derivative in terms of $y(t)$ and $y'(t)$ is displayed and the value of the second derivative at $t = 0$ is displayed. You can then find $y^{(3)}(0)$ by differentiating the equation and then plugging in the initial conditions along with the value you just found for $y''(0)$.

Find $y^{(3)}(0)$.

```
yppp0:=subs({t=0,y(t)=0,diff(y(t),t)=1,
    diff(y(t),t,t)=ypp0},diff(yppt,t));
```

The value of the third derivative at $t = 0$ is displayed. To continue this process and calculate $y^{(n)}(0)$ for more terms it would be best to store the values as you calculate them in a sequence and write a loop to do the calculations. To do any calculations with the resulting solution you would also have to use these coefficients to form a power series in t. However, Maple will perform all these calculations for you using the **series** option for **dsolve**.

Use dsolve *to calculate a series solution.*

```
sereq:=(1-t^2)*diff(y(t),t,t)-
              2*t*diff(y(t),t)+4*y(t)=0;
sereqsol:=dsolve({sereq,y(0)=0,
              D(y)(0)=1},y(t),series);
```

A fifth degree polynomial is returned. The order of the series solution returned by the **series** option of **dsolve** is determined by the global variable, **Order**, which is set to 6 by default. If you want more terms you can change the value of **Order** before calling **dsolve**.

Calculate the 10th order series solution.

```
Order:=10;
sereq:=(1-t^2)*diff(y(t),t,t)-
              2*t*diff(y(t),t)+4*y(t)=0;
sereqsol2:=dsolve({sereq,y(0)=0,
              D(y)(0)=1}, y(t),series);
```

Now a ninth degree polynomial is returned. To extract information from the series solution, such as evaluation at a point or generation of a plot, you must first convert it to a polynomial. First note that the coefficient of $y''(t)$ in this example is equal to 0 at 1 and -1, and thus you only expect the solution to be useful in the range $-1 < t < 1$.

Comparing series solutions of different order using `plot`.

```
poly1:=convert(rhs(sereqsol),polynom);
poly2:=convert(rhs(sereqsol2),polynom);
plot({poly1,poly2},t=-2..2,-5..5);
```

A plot comparing series solutions of different orders.

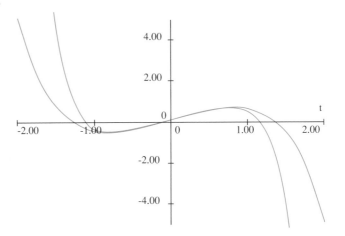

Recurrence Relations

The method of computing derivatives to calculate the terms of a Taylor series has a couple of serious drawbacks. One is that, even with Maple, computing the derivatives can be time consuming. The other is that it is difficult to find an expression for the general term of the power series. For example, the kth order term of the Taylor series for the solution of $y' = y$, $y(0) = 1$ is $x^k/k!$. This can easily be verified using the above procedure. However, it will not be so easy to do this for most equations.

The method of undetermined coefficients.

For an ordinary point, the method of undetermined coefficients does not require calculating any derivatives, but it is assumed that $a(t)$, $b(t)$, and $c(t)$ are given by polynomials in t and that $f(t) = 0$. These conditions are somewhat stronger than is necessary to apply the method. You begin by assuming there is a solution of the form

$$y(t) = \sum_{n=0}^{\infty} a_n t^n$$

where a_0 and a_1 are arbitrary. Substituting this expression into the equation

$$a(t)y''(t) + b(t)y'(t) + c(t)y(t) = 0$$

leads to conditions the coefficients a_n must satisfy. Consider the following example:

$$y'' + t^2 y = 0$$

The key to finding the coefficients a_n is that for a power series to vanish identically over any interval, each coefficient in the series must be 0. Thus, after plugging in the power series for $y(t)$, all the coefficients of t on the left-hand side of the resulting equation must be 0 since the right-hand side is 0. You can derive the following relation for this equation:

$$\sum_{n=0}^{\infty} n(n-1)a_n t^{n-2} - \sum_{n=0}^{\infty} (n^2 + 5n + 4)a_n t^n = 0.$$

By carefully extracting the coefficient of t^k and setting it equal to 0, the following relation for the coefficients is obtained:

$$a_{k+2} = -\frac{1}{(k+2)(k+1)} a_{k-2}$$

This is known as a recurrence relation. Note that to get the kth term you must do all of the sums to $k+2$ because differentiating twice reduces the order of k in the first sum by two. You sum from $k-2$ because the polynomial coefficients have order ≤ 2, thus increasing the order of k by two. To see an explanation for **coeff**, type **?coeff** at the Maple prompt.

Find recurrence relation
for a(n).

```
sereq3:=diff(y(t),t,t)+t^2*y(t)=0;
n:='n';
soly:=sum(a[n]*t^n,n=k-2..k+2);
subs(y(t)=soly,sereq3);
simplify(");
simplify(");
termk:=coeff(lhs("),t^k);
solve(termk,a[k+2]);
```

The first simplification evaluates the derivatives and the second combines the powers of t. The output of the last statement is

$$- \frac{a[k - 2]}{3 k^2 + k + 2}$$

The recurrence relation leads
to a series solution.

You can write a procedure to calculate $a(n)$ and construct a series approximation to the solution using the recurrence relation. By inspection of the output from the final statement in the previous calculation, you can see that $a[-1]$ and $a[-2]$ are needed in the calculation of $a[2]$ and $a[3]$. Note that $a_0 = y(0)$ and $a_1 = y'(0)$.

Construct the first 10 terms
of the series using the
recurrence relation.

```
a[-2]:=0;
a[-1]:=0;
a[0]:=a0;
a[1]:=a1;
for k from 0 to 7 do
   a[k+2]:=-a[k-2]/((k+2)*(k+1))
od;
k:='k';
sereq3sol:=sum(a[k]*t^k,k=0..9);
```

The ninth degree polynomial solution is displayed. You should compare this answer with the output of the command **dsolve(..., series)** on this example.

Bessel Equations

Equations with singular points.

There are series techniques to handle the case where t_0 is not an ordinary point for an equation but, in general, the resulting series are not Taylor series. Instead, they involve fractional exponents and logarithm terms. It is also likely that some solutions will not be defined at the point t_0, and in particular, you cannot usually specify initial conditions at t_0. Maple will generate series solutions to many of these examples. An example which arises in applications is the following equation known as Bessel's equation:

$$t^2 y'' + t y' + (t^2 - \mu^2) y = 0.$$

where $\mu \geq 0$ is a parameter.

Find a series solution to Bessel's equation with $\mu = 1/2$.

```
besseq:=t^2*diff(y(t),t,t)+
    t*diff(y(t),t)+(t^2-(1/2)^2)*y(t)=0;
besseqsol:=dsolve({besseq},y(t),series);
```

The series solutions to Bessel's equation have a long history and a fundamental set of solutions found by techniques beyond what can be discussed here are known as the *Bessel function of the first kind of order* μ, $J_\mu(t)$ and the *Bessel function of the second kind of order* μ, $Y_\mu(t)$. These functions are built into Maple and are returned by the **dsolve** operator when the general Bessel equation is entered.

Use dsolve to solve Bessel's equation.

```
besseqmu:=t^2*diff(y(t),t,t)+
    t*diff(y(t),t)+(t^2-(mu)^2)*y(t)=0;
besseqsol:=dsolve({besseqmu},y(t));
```
The solution is returned in terms of the built-in Maple functions, **BesselJ** and **BesselY**.

Additional Activities

1. Using paper and pencil find the recurrence relationship for

$$y' - y = 0$$

centered at $t = 0$. Following the outline of this section, use Maple to find the recurrence relationship. Since it is a first order equation, you should solve the kth term for $a[k+1]$. Does your result agree with the Taylor series of the known solution $y(t) = y(0)e^t$?

2. Find the series solution centered at $t = 0$ for the equation

$$(1 - t^2)y'' - 6ty - 4y = 0.$$

Find the recurrence relationship for the coefficients and compare the resulting series solution with the result of **dsolve, series** for the first 10 terms.

3. Find the series solution centered at $t = 0$ for the equation

$$y' + 3t^5 y = 0.$$

To find the recurrence relationship using Maple, sum from $k - 5$ to $k + 1$. Since it is a first order equation, solve the kth term for $a[k+1]$. Set $a[n] = 0$ for the appropriate number of negative values n when constructing the Taylor series from the recurrence relationship. Compare with the result of **dsolve, series** for the first 10 terms.

4. Find the series solution centered at $t = 1$ for the equation

$$y'' + ty' + 2y - t - 3 = 0, \ y(1) = 0, \ y'(1) = 1.$$

First make the substitution $t = z + 1$ and solve the equation

$$\frac{d^2 y}{dz^2} + (1 + z)\frac{dy}{dz} + 2y - 4 - z = 0$$

centered at $z = 0$ and then substitute $z = t - 1$ into the resulting Taylor series.

A Programming Under DOS

Advanced Editing

You should turn to this section when you reach the An Automating Procedure section of the book on page 27. These advanced editing features of the DOS version of Maple V allow you to easily maintain an indentation style for structured programs.

You will now create a file to use in entering the procedure on page 27.

Set up a file named mtable.

```
fred(mtable);
```

This creates a file called **mtable** and places you in the edit window where you will enter the lines of the procedure.

Enter maketable *in the edit window.*

```
maketable :=
  proc (n)
  for k from 1 by .5 to n
    do
      print(k, k^2);
    od;
  end;
```

Notice the decimal point before 5 in the third line. Be sure to use the indicated indentation as it conforms to good structured programming practice. This procedure is named **maketable**, but the file in which it is stored is named

mtable because DOS allows at most eight characters in a file name.

Saving the mtable *file.*

Press F2 to save the procedure in the **mtable** *file. Press F3 to read this procedure into your Maple session window.* Notice that the program is displayed on the screen.

Let's use this procedure.

```
maketable(5);
```

As you might expect, a table of squares is displayed.

Let's generalize the **maketable** procedure to allow for a variable increment.

Bring the maketable *procedure into the edit window.*

```
fred(mtable);
```

Notice that the cursor is at the first character of the procedure name.

Change the second and third lines of the procedure definition as follows.

```
maketable :=
    proc (n, increment)
    for k from 1 by increment to n
        do
            print(k, k^2);
        od;
    end;
```

Use the arrow keys to move to the appropriate places to insert and delete characters.

Save the edited procedure and exit the edit window.

Press F2 and then F3.
Notice that the new procedure is displayed on the screen.

Let's use this new procedure.

```
maketable(3, .2);
```

Are the values displayed what you would expect? The step size is now .2 and the last pair is 3, 9.

You can use this same method for the remaining programs on pages 28, 29, and 30 as well as for any other programming you may do.

Index

Accessing array components, 133, 135
Accessing list components, 133
Accessing matrix entries, 68
Addition, 4
addrow, 71
adjoint, 87, 88
Adjoint of a matrix, 87, 88
alias, 81, 87, 88, 111
allvalues, 106, 107, 111, 150
angle, 90
Applications, 53, 61, 121
Approximations, 6, 16, 17, 138, 145
Argument, 68, 69, 82
array, 85, 133
Array, 67, 69, 80, 133, 135
Arrow keys, 10, 12, 13
Assignment statement, 6-8, 21, 24
Asymptote, 22-23, 34-35
augment, 73, 75, 89, 101
Augmenting a matrix, 70, 73, 75
Automating commands, 26-30
Autonomous differential equation, 155
Axes Style menu, 60

Back substitution, 72
Backspace key, 10
backsub, 72, 74
basis, 96-98
Basis, 96, 101
Bessel equation, 174
Boxed, 60
Braces, 20, 70, 120
Brackets, 17, 32, 68, 90
BlockDiagonal, 85

Caret, 4
Cayle-Hamilton theorem, 91, 152
Center of mass, 61

changevar, 52
Characteristic matrix, 105, 113
Characteristic polynomial, 91, 92, 112, 151
Characteristic value, 104
Characteristic vector, 104
charmat, 105, 115
charpoly, 91, 105
Checking results, 8, 15
Close box, 4, 8, 9
cmpltsq, 16
coeff, 172, 173
col, 74
Colon, 29, 30, 133
Color Patch, 60
colspace, 96
colspan, 98
Command menu, 12
Commands, 6, 7, 21
companion, 85, 105
completesquare, 16
Complex eigenvalue, 150
Complex eigenvector, 150
Complex solutions, 16, 17
Concatenation, 90
Consistent system, 73
Constrained, 60
convert, 57, 100, 171
coords, 33
copy, 75
Copy, 10
copyinto, 76
cos, 18, 53
Cross product, 89
crossprod, 89
curl, 59
Curl of a function, 59
Cursor, 9
Curve fitting, 92

D, 118, 144, 145
Decimal numbers, 6
Defining functions, 24, 26, 30
Defining procedures, 25
Definite integrals, 51
Delete, 11, 74
delcols, 74
delrows, 74
denom, 21, 22
Derivatives, 48, 49, 54, 57
det, 41, 87
Determinant of a matrix, 41, 87
diag, 82, 83
Diagonal matrix, 82, 83, 109, 110
diff, 48, 49, 118
Differential equations, 117, 118
Digits, 6
Dimension, 40, 68, 97, 98, 148
directionfield, 127, 128
Directionfield options, 128
Direction field, 126, 127
display, 130, 136, 157
Displaying several graphs, 31
diverge, 59
Divergence of a vector, 59
Division, 4
do/od, 26, 27, 83
Domain, 9, 32, 59
DOS, 2, 4, 11, 13, 15, 27, 30, 49, 59, 177
Dot product, 89
dotprod, 89
Double integrals, 61
dsolve, 118, 119, 164

Edit menu, 10
Editing, 10, 11
Editing matrices, 69, 74, 84
Eigenspace, 105, 107, 111
eigenvals, 105, 107
Eigenvalues, 104, 151
Eigenvectors, 104, 109, 151
eigenvects, 107, 110, 149
else, 25
end, 25,
Enter key, 2-4, 27, 28
Entering expressions, 3, 7
Entries of a matrix, 68, 69
Equations
    Differential, 117, 118, 144
    Exponential, 19
    Logarithmic, 19
    Polynomial, 15, 37

    Trigonometric, 18
    Systems of, 20, 69, 73, 88, 145, 154
Equilibrium point, 158
Error estimation, 138, 140
Errors, 5-7, 79
Esc, 4, 31
Euler's method, 132, 134, 137
evalc, 107, 108
evalf, 6, 17, 123, 133
evalm, 40, 68, 77
Evaluating expressions, 7
Evaluating functions, 24
Existence of a solution, 121, 141, 147
exp, 52, 56, 149
expand, 8, 53
explicit, 120, 164
Explicit solution, 120, 122
exponential, 152, 153
Exponential equations, 19
Exponentiation, 4, 5
Expression, 6, 7

factor, 8, 38
ffgausselim, 72
fi, 25
File menu, 11
First order differential equations, 118, 159
firsteuler, 138, 139
Floating point, 6
Flow lines. 129, 131
for, 26-30, 83
For loop, 26-30, 83, 133
Format menu, 13
frobenius, 114, 115
Frobenius form, 114, 115
fsolve, 16-18, 19, 121
Function behavior, 34-36, 38, 47, 48
Function keys, 12, 13
Function values, 24-26
Functions of one variable, 24, 34
Functions of several variables, 26, 57

Gauss-Jordan elimination, 72
gausselim, 71, 73
Gaussian elimination, 71, 72
gaussjord, 72
General solution, 121, 149, 152
genmatrix, 100
Go Away box, 4, 8
GramSchmidt, 96
Graphing, 9, 31
grid, 128, 157

Heaviside, 166, 167
Help facility, 3, 11, 67
Higher order derivatives, 49, 54, 57
Higher order differential equations, 144, 160
Highlight, 10, 27

IBM, 2, 4, 11, 12, 27, 117
Identity matrix, 41, 81, 82
if/fi, 25
If Then Else, 25
Im, 150, 151
impeuler, 138, 139
Implicit solution, 120
Inequalities, 19
Initial value problem, 120, 138, 141, 145
Inner product, 77, 80, 89
innerprod, 77, 80 89
int, 50, 51, 61, 142
Int, 51, 52
intbasis, 98
Integer numbers, 5
Integration, 50-52, 61
intparts, 51, 52
inverse, 41, 88
Inverse of a matrix, 41
Inverse Laplace transform, 165
invlaplace, 165
Irrational numbers, 15
Isoclines, 129-131
isolate, 16
iterations, 128, 129, 157

kernal, 100
Kernal, 97, 100

laplace, 164, 165
Laplace transforms, 164, 165
leastsqrs, 102
Left-hand limits, 47
limit, 44-47, 62
Limits, 45-47, 62
Limits at infinity, 46
Line terminator, 4
Linear algebra, 67
Linear transformation, 99, 100
Linearly independent, 97, 109, 149, 151
linsolve, 88, 89, 93, 97
List, 58, 68, 82, 133
List-of-lists, 68, 81
ln, 19, 50

Logarithmic equations, 19
Loops, 26-30, 83, 133

Macintosh, 1, 10
makelist, 134, 161
map, 92
Maple, 1
Maple built-in functions
    cos, 18, 53
    exp, 52, 56, 149
    Im, 150, 151
    ln, 19, 50
    Re, 150, 151
    sec, 50
    seq, 90, 91
    sin, 18, 51, 52
    sqrt, 50, 54
Maple commands
    addrow, 71
    adjoint, 87, 88
    alias, 81, 87, 88, 111
    allvalues, 106, 107, 111, 150
    array, 85
    augment, 73, 75, 89, 101
    backsub, 72, 74
    basis, 96-98
    BlockDiagonal, 85
    Boxed, 60
    changevar, 52
    charmat, 105, 115
    charpoly, 91, 105
    coeff, 172, 173
    col, 74
    colspace, 96
    colspan, 98
    companion, 85, 105
    completesquare, 16
    Constrained, 60
    convert, 57, 100, 171
    coords, 33
    copy, 75
    copyinto, 76
    curl, 59
    D, 144
    delcols, 74
    delrows, 74
    denom, 21, 22
    det, 41, 87
    diag, 82, 83
    diff, 48, 49, 118

Maple commands (continued)

directionfield, 127

display, 130, 136, 157

diverge, 59

do/od, 26, 27, 83

dotprod, 89

dsolve, 118, 119, 164

eigenvals, 105, 107

eigenvects, 107, 110, 149

else, 25

end, 25

evalc, 107, 108

evalf, 6, 17, 123, 133

evalm, 40, 68, 77

expand, 8, 53

explicit, 120, 164

exponential, 152, 153

factor, 8, 38

ffgausselim, 72

fi, 25

firsteuler, 138, 139

for, 26-30, 83

frobenius, 114, 115

fsolve, 16-18, 19, 121

gausselim, 71, 73

gaussjord, 72

genmatrix, 100

GramSchmidt, 96

grid, 128, 157

Heaviside, 166, 167

if/fi, 25

impeuler, 138, 139

innerprod, 77, 80, 89

int, 50, 51, 61, 142

Int, 51, 52

intbasis, 98

intparts, 51, 52

inverse, 41, 88

invlaplace, 165

isolate, 16

iterations, 128, 129, 157

kernal, 100

laplace, 164, 165

leastsqrs, 102

limit, 45-47, 62

linsolve, 88, 89, 93, 97

makelist, 134, 161

map, 92

matrix, 40, 68, 81, 149, 153

minor, 76

mulrow, 71

norm, 89, 97

NULL, 28

nullspace, 97, 105

numer, 21, 22

numeric, 164

numsteps, 157, 158

od, 26, 27, 83

op, 106

phaseplot, 156, 157

picard, 142, 143

pivot, 71

plot, 9, 10, 18, 20, 31, 119

plot3d, 26, 59, 60

POINT, 22, 25

polar, 33

polynom, 57, 171

powsubs, 52, 53

print, 26, 79, 133

proc, 25, 27, 47, 99

quo, 35

randmatrix, 83

range, 97

rank, 97

read, 117

readlib, 166

rem, 92

rhs, 119, 146

RootOf, 106, 107, 150

row, 74, 91, 96

rowspace, 96

rowspan, 98

rref, 72

rungekutta, 138

rungekuttahf, 140, 161

seq, 90, 91

series, 164, 170, 173, 174

simplify, 21, 22, 49, 124

smith, 114

solve, 9, 15, 16, 19, 73

spacecurve, 146, 155

sparse, 83

stack, 75

stepsize, 128, 129, 157

submatrix, 73, 75

subs, 15, 20, 91, 119, 123

subvector, 76

sum, 56, 112, 167

sumbasis, 98

swapcol, 41

swaprow, 41, 71

taylor, 56

transpose, 41, 89, 114

value, 53

vandermonde, 93
vector, 69, 73, 81, 99
view, 59
with, 16, 40, 130
Maple library, 117, 166
Maple packages
   linalg, 40, 58, 67
   plots, 130
   student, 16, 63
Maple session window, 12, 49, 117, 127, 166
Maple statement length, 25
Mapping arrow, 24, 26
matrix, 40, 81, 149, 153
Matrix, 40-42, 59, 68
Matrix
   Addition, 40, 80
   Adjoint, 87
   Assigning, 40, 68
   Augmenting, 70, 73
   Determinant, 41
   Diagonal, 82, 110, 111
   Dimensions, 40, 68
   Display, 40, 68
   Editing, 69, 74
   Entries, 68, 69
   Identity, 41, 81, 82
   Inverse, 41, 88
   Multiplication, 41, 77, 78
   Rank, 97
   Subscripts, 68-70
   Symmetric, 110
   Transpose, 41
Method of undetermined coefficients, 145, 171
Minima, 49
minor, 76
Mouse button, 1, 10
mulrow, 71
Multiple integrals, 61
Multiplication, 7
Multiplication of matrices, 41, 77, 78

norm, 89, 97
Norm of a vector, 89
NULL, 28
nullspace, 97, 105
Numbers, 5
numer, 21, 22
numeric, 164
numsteps, 157, 158

od, 26, 27, 83
ODE, 117, 127, 137, 142

ODE.m, 117
op, 106
Orbit, 156-158
Order, 170
Order term, 56
Ordinary point, 169, 171
Orthogonal basis, 96
Orthogonal projection, 101

Parametric form, 32
Parentheses, 5, 22
Partial derivatives, 57, 58
Paste, 10
Patch, 60
Phase portrait, 156, 158
Phase space, 154, 156
Phase vector, 155, 156
phaseplot, 156, 157
Pi, 18
picard, 142, 143
Picard iterates, 141, 142
pivot, 71
plot, 9, 10, 18, 20, 31, 119
Plot Style menu, 60
Plot window, 9
plot3d, 26, 59, 60
Plotting in three-dimensions, 26, 59
Plotting sets of functions, 20, 31
POINT, 22, 25
Points of inflection, 49
polar, 33
Polar form, 33
polynom, 57, 171
Polynomial equations, 15, 37
powsubs, 52, 53
print, 26, 79, 133
Printing, 12, 13
proc, 25, 27, 47, 99
Procedure, 25, 27, 47, 99
Programming, 26-30, 134
Prompt, 1, 2

Quit, 11
quo, 35
Quote symbols
   Backward, 117, 118
   Double, 8, 16
   Single, 7, 84, 90

randmatrix, 83
range, 97
Range, 9, 32, 59

rank, 97
Rational expressions, 21, 22
Rational numbers, 5
Re, 150, 151
read, 117
readlib, 166
Recurrence relations, 171, 172
Reestablishing variables, 7
rem, 92
Resizing the Plot window, 33
Return key, 3, 4, 27
rhs, 119, 146
Right-hand limits, 47
RootOf, 38, 106, 107, 150
Roots of polynomials, 36
row, 74, 91, 96
Row reduced echelon form, 72
rowspace, 96
rowspan, 98
rref, 72
rungekutta, 138
Runge/Kutta, 137, 159
rungekuttahf, 140, 161

Save, 12
Saving, 12, 13
Scale Type menu, 60
sec, 50
seq, 90, 91
Sequence of solutions, 17
Sequences, 90
series, 164, 170, 173, 174
Series, 56
Series solutions, 151, 169
Session menu, 13
Set notation, 19
Sets, 19
Similar matrices, 113
simplify, 21, 22, 49, 124
sin, 18, 51, 52
Singular point, 174
Size box, 33
smith, 114
Smith form, 113
solve, 9, 15, 16, 19, 73
Solving equations, 9, 15, 16
Solving for a variable, 16
spacecurve, 146, 155
Span, 96, 101, 149
sparse, 83
sqrt, 50, 54
stack, 75

stepsize, 128, 129, 157
Style, 22, 25
submatrix, 73, 75
subs, 15, 20, 91, 119, 123
Subscripts, 68-70, 90
Subspace, 97, 98, 101
Subtraction, 8
subvector, 76
sum, 56, 112, 167
sumbasis, 98
Sun Workstation, 3, 27, 59, 117
swapcol, 41
swaprow, 41, 71
Symbolic matrix, 83, 87
Symbolic vector, 83, 87
Symmetric matrix, 110
Syntax error, 6
Systems of differential equations, 145, 146, 154
Systems of equations, 20, 69, 73, 88

taylor, 56
Taylor polynomial, 169
Taylor series, 56, 171
Terminator, 4
Title bar, 33
transpose, 41, 89, 114
Transpose of a matrix, 41
Trigonometric equations, 18
Types of numbers, 5

Unaliasing a variable, 108
Unassigning a function definition, 26
Unassigning a variable, 7, 84
Uniqueness of a solution, 121, 141, 165

value, 53
vandermonde, 93
Vandermonde matrices, 93
Variable, 7
vector, 69, 73, 81, 99
Vector, 58, 68, 69
Vector function, 58, 68, 69
Vector space, 96, 148
Verifying solutions, 124, 125
view, 59

Windows, 1, 2, 4, 11, 12, 27, 59
Windows menu, 11
with(linalg), 40, 58, 67, 149
with(plots), 130, 146
with(student), 16, 51

Zeros of polynomials, 36

@, 144, 145
', 7, 84, 90
`, 117, 118
", 8, 16
*, 7
&*, 77, 78
/, 4
%, 17, 112, 124
=, 9
>, 25
., 90
;, 4
:, 29, 30, 133
:=, 6-8
?, 3
^, 4
->, 24, 26
π, 18, 19
⌘, 10

Performance, reliability, and the most power for your dollar—it's all yours with Maple V!

Maple V Release 3 Student Edition *for Macintosh or DOS/Windows* Just $84.50.

Offering numeric computation, symbolic computation, graphics, and programming, **Maple V Release 3 Student Edition** gives you the power to explore and solve a tremendous range of problems with unsurpassed speed and accuracy. Featuring both 3-D and 2-D graphics and more than 2500 built-in functions, **Release 3** offers all the power and capability you'll need for the entire array of undergraduate courses in mathematics, science, and engineering.

Maple V's *vast* library of functions also provides sophisticated scientific visualization, programming, and document preparation capabilities, **including the ability to output standard mathematical notation.**

With Release 3, you can:

- plot implicit equations in 2-D and 3-D

- generate contour plots

- apply lighting (or shading) models to 3-D plots and assign user-specified colors to each plotted 2-D function

- view 2-D and 3-D graphs interactively and use Release 3's animation capabilities to study time-variant data

- view and print documents with standard mathematical notation for Maple output (including properly placed superscripts, integral and summation signs of typeset quality, matrices, and more)

- save the state (both mathematical and visual) of a Maple session at any point—and later resume work right where you left off

- migrate Maple worksheets easily across platforms. (This is especially valuable for students using Maple V on a workstation in a computer lab who then want to continue work on their own personal computers)

- export to LaTeX and save entire worksheets for inclusion in a publication-quality document *(New!)*

Linear Algebra Commands

augment
p. 73

Join two or more matrices together horizontally.

```
augment(matrix([[3,7],[8,1]],
matrix([[5],[4]]);
```

backsub
p. 72

Back substitution on a matrix.

```
A:=matrix([[5,-2,3,-1],
    [0,7,-1,2],[0,0,-4,8]]);
backsub(A);
```

charpoly
p. 91

Computes the characteristic polynomial of a square matrix.

```
A:=matrix(3,3,(i,j)->i+j-1
    mod 3);
charpoly(A,x);
```

convert
p. 100

Converts a series to a polynomial.

```
convert(seriesone,polynom);
```

det
p. 41

Computes the determinant of a matrix.

```
A:=matrix([[-3,5],[8,-1]]);
det(A);
```

diag
p. 82

Creates a diagonal matrix.

```
diag(1,5,2,4,3); or
diag(Z,1,2,matrix(3,3,1));
```

eigenvals
p. 105

Computes the eigenvalues of a matrix.

```
A:=matrix([[2,2],[3,1]]);
eigenvals(A);
```

evalm
p. 68

Displays or evaluates a matrix.

```
A:=matrix([[3,-4,7],[2,7,1]]);
evalm(A);
```

gausselim
p. 71

Performs fraction-free Gaussian elimination on a matrix.

```
A:=matrix([[3,-4,7],[2,7,1]]);
gausselim(A);
```

genmatrix
p. 100

Generates the coefficient matrix from a set of equations.

```
genmatrix([x+2*y=3,3*x-5*y=0],
[x,y]);
```

inverse
p. 88

Computes the inverse of a matrix.

```
A:=matrix([[3,2,4],[4,-2,6],
    [8,3,5]]);
inverse(A);
```

linsolve
p. 88

Returns the solution of a system of linear equations.

```
A:=matrix([[1,-1,4],[3,2,-1],
    [2,1,-1]]);
v:=vector([-2,3,1]);
linsolve(A,v);
```

matrix
p. 68

Creates a matrix.

```
A:=matrix([[5,1],[-3,6]]); or
A:=matrix(2,2,[5,1,-3,6]);
```

transpose
p. 89

Computes the transpose of a matrix.

```
A:=matrix([[5,1,0],[-3,6,1],
    [-2,5,4]]);
transpose(A);
```

vector
p. 69

Creates a vector.

```
v:=vector([1,2,3]);
```

Differential Equation Commands

directionfield
p. 127

Plots the direction field for a differential equation.

```
de:=(t,y) -> y^2+t^2-1;
fieldplot(de,-2..2,-2..2);
```

dsolve
p. 118

Solves differential equations.

```
dsolve({diff(v(t),t)+2*t=0,
    v(1)=5},v(t));
```